Semantics and t

I0059616

Solution Man

**Semantics
and the Syntax of Algebra**
Solution Manual
Afshin Azari-Vala

Copyright © 2018 Afshin Azari-Vala

All rights reserved. No part of this publication may be reproduced or transmitted in any form by any means, electronic or mechanical, including photocopying and recording, or by any information storage or retrieval system, without written permission from the author.

Cover design by Afshin Azari-Vala

Published by Afshin Azari-Vala

To contact the publisher please send email to admin@tersemath.com.

ISBN 978-1-7750996-1-1

Disclaimer: Every effort has been made to make sure that this publication is free of error. However, the author and the publisher do not assume and hereby disclaim any liability to any party for any loss, damage, or disruption caused by any errors, typographical or other, that may have crept into this publication.

An Overview of the Manual

This publication provides detailed solutions for the problems in the exercise sets in the textbook *Semantics and the Syntax of Algebra* by the author. The coverage includes both even-numbered and odd-numbered problems in the exercise sets.

As our aim is to promote formulations and algorithms that promote fluency, for the most part, we have provided detailed solutions as opposed to partial solutions or just answers. Alternative approaches that meet the criteria for fostering fluency in solving mathematical problems are included in the solutions.

When an exercise extends a concept that is introduced in the body of the textbook or introduces a new one, further detail about the topic is included in the form of side notes. This should help the reader connect the ideas that are presented in the body of the text to their extensions in the exercises.

Ideally, this manual should be consulted as a check on solutions that the reader provides. While study skills differ, it is often the case that active engagement through solving problems *using pen and paper* is much more effective than passive participation through looking up the solutions to problems that are posed for practice.

Solutions to Exercise Sets

Exercise Set 1.1

1. work, W, joule, J
2. number of, n, one, 1
3. resistance, R, ohm, Ω
4. mass, m, gram, g
5. volume, V, litre, L
6. force, F, newton, N
7. number of, n, one, 1
8. number of, n, one, 1
9. amount, a, dollar, $

10. pressure, p, pascal, Pa
11. number of, n, one, 1
12. amount, a, dollar, $
13. speed, s,[1] metres per second, m/s
14. length, l, centimetre, cm
15. rate, r, one, 1
16. mass, m, atomic mass unit, amu
17. temperature, t, degree Celsius, °C
18. area, A, square metre, m^2

Exercise Set 1.2.1

1. 1. Calculate the amount of money spent on the pens.

   $$4.20 \times 3 = 12.60$$

   I spent $12.60 on the pens.

   2. Calculate the amount of money spent on the notebooks.

   $$6.99 \times 5 = 34.95$$

   I spent $34.95 on the notebooks.

   3. Calculate the total amount of money spent.

   $$12.60 + 34.95 = 47.55$$

   I spent $47.55 in total.

---

[1] As explained in Appendix A, the symbol v may also be used to represent the value of the quantity *speed*.

1

2. 1. Calculate the amount of money the bookstore receives from the sale of the used textbooks.

$$42.50 \times 97 = 4122.50$$

The bookstore receives $4122.50 from the sale of the used textbooks.

2. Calculate the amount of money the bookstore receives from the sale of the new textbooks.

$$75.99 \times 54 = 4103.46$$

The bookstore receives $4103.46 from the sale of the new textbooks.

3. Calculate the amount of money the bookstore receives from the sale of all the textbooks.

$$4122.50 + 4103.46 = 8225.96$$

The bookstore receives $8225.96 from the sale of all the textbooks.

3. 1. Calculate the energy contribution from the carbohydrate content of the cup of yogurt.

$$4 \times 6.2 = 24.8$$

Carbohydrates contribute 24.8 Cal to the total energy content of the cup of yogurt.

2. Calculate the energy contribution from the fat content of the cup of yogurt.

$$9 \times 8.5 = 76.5$$

Fats contribute 76.5 Cal to the total energy content of the cup of yogurt.

3. Calculate the energy contribution from the protein content of the cup of yogurt.

$$4 \times 13.7 = 54.8$$

Proteins contribute 54.8 Cal to the total energy content of the cup of yogurt.

4. Calculate the energy content of the cup of yogurt.

$$24.8 + 76.5 + 54.8 = 156.1$$

The cup of yogurt contains 156.1 Cal of energy.

4. 1. Calculate Aisha's intake of energy from the cups of coffee.

   $30 \times 3 = 90$

   Aisha had an intake of energy of 90 Cal from the cups of coffee.

2. Calculate Aisha's intake of energy from the servings of fruit.

   $110 \times 2 = 220$

   Aisha had an intake of energy of 220 Cal from the servings of fruit.

3. Calculate Aisha's loss of energy due to working.

   $130 \times 2 = 260$

   Aisha had a loss of energy of 260 Cal due to working.

4. Calculate Aisha's net intake of energy this afternoon.

   $90 + 220 + 87 - 260 = 137$

   Aisha had a net energy intake of 137 Cal this afternoon.

5. 1. Calculate the cost of mailing the letters.

   $2.50 \times 18 = 45$

   It will cost $45 to mail the letters.

2. Calculate the cost of mailing the packages.

   $8.10 \times 12 = 97.20$

   It will cost $97.20 to mail the packages.

3. Calculate the cost of mailing the boxes.

   $19.50 \times 8 = 156$

   It will cost $156 to mail the boxes.

4. Calculate the cost of mailing the letters, packages and boxes.

   $45 + 97.20 + 156 = 298.20$

   It will cost $298.20 to mail the letters, packages and boxes.

6. 1. Calculate Joan's earnings from clerical work this week.

   $16.35 \times 8 = 130.80$

   Joan earned $130.80 from her clerical work this week.

2. Calculate Joan's earnings from her waitressing job this week.

$$15.20 \times 4.5 = 68.40$$

Joan made $68.40 from her waitressing job this week.

3. Calculate Joan's earnings from her fundraising job this week.

$$12.50 \times 12 = 150$$

Joan made $150 from her fundraising job this week.

4. Calculate Joan's total earnings this week.

$$130.80 + 68.40 + 150 = 349.20$$

Joan earned $349.20 this week.

7. 1. Calculate the mass of N atoms in a molecule of N_2O_3.

$$14.01 \times 2 = 28.02$$

The N atoms have a mass of 28.02 amu.

2. Calculate the mass of O atoms in a molecule of N_2O_3.

$$16.00 \times 3 = 48$$

The O atoms have a mass of 48 amu.

3. Calculate the mass of a molecule of N_2O_3.

$$28.02 + 48 = 76.02$$

The mass of a molecule of N_2O_3 is 76.02 amu.

8. 1. Calculate distance covered this morning.

$$110 \times 2.5 = 275$$

I covered a distance of 275 km this morning.

2. Calculate distance covered this afternoon.

$$90 \times 1.5 = 135$$

I covered a distance of 135 km this afternoon.

3. Calculate total distance covered today.

$$275 + 135 = 410$$

I covered a distance of 410 km today.

9. 1. Calculate the mass of recycling material that arrived at the recycling plant during the time period mentioned.

$$17.5 \times 2.75 = 48.125$$

48.125 t of recycling material entered the plant during the time period mentioned.

2. Calculate the mass of recycling material processed at the plant during the time period mentioned.

$$14 \times 1.5 = 21$$

The plant has processed 21 t of recycling material during the time period mentioned.

3. Calculate the mass of recycling material waiting to be processed.

$$48.125 - 21 = 27.125$$

27.125 t of recycling material is waiting to be processed.

10. 1. Calculate distance covered during the three days of cycling.

   a. Calculate distance covered during the first day of cycling.

$$20 \times 0.75 = 15$$

   I covered a distance of 15 km during the first day of cycling.
   b. Calculate distance covered during the second day of cycling.

$$18.5 \times 1.5 = 27.75$$

   I covered a distance of 27.75 km during the second day of cycling.
   c. Calculate distance covered during the third day of cycling.

$$23.5 \times 0.5 = 11.75$$

   I covered a distance of 11.75 km during the third day of cycling.
   d. Calculate distance covered during the three days of cycling.

$$15 + 27.75 + 11.75 = 54.5$$

   I covered a distance of 54.5 km during the three days of cycling.

2. Calculate additional distance to cover to meet goal.

$$100 - 54.5 = 45.5$$

To meet my goal, I need to cover an additional distance of 45.5 km this week.

Exercise Set 1.2.2

1.

a amount of money I spent ($)
$a = 4.20 \times 3 + 6.99 \times 5$

$a = 12.60 + 34.95$
$a = 47.55$

I spent $47.55.

2.

a amount of money received from the sale of the textbooks ($)
$a = 42.50 \times 97 + 75.99 \times 54$

$a = 4122.50 + 4103.46$
$a = 8225.96$

The bookstore receives $8225.96 from the sale of the textbooks.

3.

E Energy content of the cup of yogurt (Cal)
$E = 4 \times 6.2 + 9 \times 8.5 + 4 \times 13.7$

$E = 24.8 + 76.5 + 54.8$
$E = 156.1$

The cup of yogurt contains 156.1 Cal of energy.

4.

E Aisha's net intake of energy this afternoon (Cal)
$E = 30 \times 3 + 110 \times 2 + 87 - 130 \times 2$

$E = 90 + 220 + 87 - 260$
$E = 137$

Aisha had a net intake of energy of 137 Cal this afternoon.

5.

c the cost of mailing the letters, packages, and boxes ($)
$c = 2.50 \times 18 + 8.10 \times 12 + 19.50 \times 8$

$c = 45 + 97.20 + 156$
$c = 298.20$

It costs $298.20 to mail the letters, packages, and boxes.

6.

e Joan's earnings this week ($)
$e = 16.35 \times 8 + 15.20 \times 4.5 + 12.50 \times 12$

$e = 130.80 + 68.40 + 150$
$e = 349.20$

Joan earned $349.20 this week.

7.

m mass of a molecule of N_2O_3 (amu)
$m = 14.01 \times 2 + 16.00 \times 3$

$m = 28.02 + 48$
$m = 76.02$

A molecule of N_2O_3 has a mass of 76.02 amu.

8.

d total distance I covered today (km)
$d = 110 \times 2.5 + 90 \times 1.5$

$d = 275 + 135$
$d = 410$

In total, I covered a distance of 410 km today.

9.

m mass of recycling material waiting to be processed (t)
$m = 17.5 \times 2.75 - 14 \times 1.5$

$m = 48.125 - 21$
$m = 27.125$

27.125 t of recycling material is waiting to be processed.

10.

d additional distance to cover to meet goal (km)
$d = 100 - (20 \times 0.75 + 18.5 \times 1.5 + 23.5 \times 0.5)$

$d = 100 - (15 + 27.75 + 11.75)$
$d = 100 - 54.5$
$d = 45.5$

I need to cover an additional distance of 45.5 km to meet my goal.

Exercise Set 2.1

1. a. **A Step-by-Step Solution**

1. Calculate the amount of money spent on the pens.

$$2.50 \times 6 = 15$$

I spent $15 on the pens.

2. Calculate the amount of money spent on the notebooks.

$$4.99 \times 8 = 39.92$$

I spent $39.92 on the notebooks.

3. Calculate total amount of money spent.

$$15 + 39.92 = 54.92$$

In total I spent $54.92.

An Algebraic Solution

a amount of money spent ($)
$a = 2.50 \times 6 + 4.99 \times 8$

$a = 15 + 39.92$
$a = 54.92$

In total I spent $54.92.

b. A Step-by-Step Solution

1. Calculate the amount of money spent on the fabric.

$$1.20 \times 6.5 = 7.80$$

I spent $7.80 on the fabric.

2. Calculate the amount of money spent on the rolls of string.

$$2 \times 3 = 6$$

I spent $6.00 on the rolls of string.

3. Calculate total amount of money spent.

$$7.80 + 6.00 = 13.80$$

In total I spent $13.80.

An Algebraic Solution

a amount of money spent ($)
$a = 1.20 \times 6.5 + 2 \times 3$

$a = 7.80 + 6$
$a = 13.80$

In total I spent $13.80.

c. A Step-by-Step Solution

1. Calculate the number of packages of printer paper.

$$18 \times 20 = 360$$

There are 360 packages of printer paper.

2. Calculate the cost of the order.

$$2.99 \times 360 = 1076.40$$

The order cost $1076.40.

An Algebraic Solution

c cost of the order ($)
$c = 2.99 \times 18 \times 20$

$c = 1076.40$
The order cost $1076.40.

d. **A Step-by-Step Solution**

1. Calculate the cost of ordering the surgical gloves.

$$5.20 \times 20 = 104$$

The surgical gloves cost $104.

2. Calculate the cost of ordering the masks.

$$10.99 \times 32 = 351.68$$

The masks cost $351.68.

3. Calculate the cost of the order.

$$104 + 351.68 = 455.68$$

The order cost $455.68.

An Algebraic Solution

c cost of the order ($)
$c = 5.20 \times 20 + 10.99 \times 32$

$c = 104 + 351.68$
$c = 455.68$
The order cost $455.68.

e. **A Step-by-Step Solution**

1. Calculate volume of medication used.

$$5.65 \times 12 = 67.8$$

I used 67.8 mL of medication.

2. Calculate volume of medication left.

$$400 - 67.8 = 332.2$$

I have 332.2 mL of medication left.

An Algebraic Solution

V volume of medication left (mL)
$V = 400 - 5.65 \times 12$

$V = 400 - 67.8$
$V = 332.2$

I have 332.2 mL of medication left.

f. A Step-by-Step Solution

1. Calculate the volume of the salt solution used.

$$2.4 \times 3 = 7.2$$

I used 7.2 L of the salt solution.

2. Calculate the volume of the salt solution left.

$$180 - 7.2 = 172.8$$

I have 172.8 L of the salt solution left.

An Algebraic Solution

V volume of the salt solution left (L)
$V = 180 - 2.4 \times 3$

$V = 180 - 7.2$
$V = 172.8$

I have 172.8 L of the salt solution left.

g. A Step-by-Step Solution

1. Calculate the mass of the C atoms.

$$12.01 \times 2 = 24.02$$

The C atoms have a mass of 24.02 amu.

2. Calculate the mass of the H atoms.

$$1.008 \times 6 = 6.048$$

The H atoms have a mass of 6.048 amu.

3. Calculate the mass of a molecule of C_2H_6.

$$24.02 + 6.048 = 30.068$$

A molecule of C_2H_6 has a mass of 30.068 amu.

An Algebraic Solution

m mass of a molecule of C_2H_6 (amu)
$m = 12.01 \times 2 + 1.008 \times 6$

$m = 24.02 + 6.048$
$m = 30.068$

A molecule of C_2H_6 has a mass of 30.068 amu.

h. **A Step-by-Step Solution**

1. Calculate the mass of the C atoms.

$$12.01 \times 2 = 24.02$$

The C atoms have a mass of 24.02 g.

2. Calculate the mass of the H atoms.

$$1.008 \times 6 = 6.048$$

The H atoms have a mass of 6.048 g.

3. Calculate the mass of 1 mol of C_2H_6.

$$24.02 + 6.048 = 30.068$$

The mass of 1 mol of C_2H_6 is 30.068 g.

An Algebraic Solution

m mass of 1 mol of C_2H_6 (g)
$m = 12.01 \times 2 + 1.008 \times 6$

$m = 24.02 + 6.048$
$m = 30.068$

The mass of 1 mol of C_2H_6 is 30.068 g.

i. **A Step-by-Step Solution**

1. Calculate the mass the O atoms.

$$16.00 \times 2 = 32$$

The O atoms have a mass of 32 g.

2. Calculate the mass of 1 mol of SO_2.

$$32.07 + 32 = 64.07$$

The mass of 1 mol of SO_2 is 64.07 g.

An Algebraic Solution

m mass of 1 mol of SO_2 (g)
$m = 32.07 + 16.00 \times 2$

$m = 32.07 + 32$
$m = 64.07$

The mass of 1 mol of SO_2 is 64.07 g.

j. A Step-by-Step Solution

1. Calculate the mass the H atoms.

$$1.008 \times 4 = 4.032$$

The H atoms have a mass of 4.032 amu.

2. Calculate the mass of a molecule of CH_4.

$$12.01 + 4.032 = 16.042$$

A molecule of CH_4 has a mass of 16.042 amu.

An Algebraic Solution

m mass of a molecule of CH_4 (amu)
$m = 12.01 + 1.008 \times 4$

$m = 12.01 + 4.032$
$m = 16.042$

A molecule of CH_4 has a mass of 16.042 amu.

k. A Step-by-Step Solution

1. Calculate the mass the H atoms.

$$1.008 \times 2 = 2.016$$

The H atoms have a mass of 2.016 amu.

2. Calculate the mass of a molecule of H_2O.

$$2.016 + 16.00 = 18.016$$

A molecule of H_2O has a mass of 18.016 amu.

An Algebraic Solution

m mass of a molecule of H_2O (amu)
$m = 1.008 \times 2 + 16.00$

$m = 2.016 + 16$
$m = 18.016$

A molecule of H_2O has a mass of 18.016 amu.

l. **A Step-by-Step Solution**

1. Calculate the mass the O atoms.

$$16.00 \times 2 = 32$$

The O atoms have a mass of 32 g.

2. Calculate the mass of 1 mol of CO_2.

$$12.01 + 32 = 44.01$$

The mass of 1 mol of CO_2 is 44.01 g.

An Algebraic Solution

m mass of 1 mol of CO_2 (g)
$m = 12.01 + 16.00 \times 2$

$m = 12.01 + 32$
$m = 44.01$

The mass of 1 mol of CO_2 is 44.01 g.

m. **A Step-by-Step Solution**

1. Calculate the mass of 1 mol of O_3.

$$16.00 \times 3 = 48$$

The mass of 1 mol of O_3 is 48 g.

An Algebraic Solution

m mass of 1 mol of O_3 (g)
$m = 16.00 \times 3$

$m = 48$

The mass of 1 mol of O_3 is 48 g.

n. **A Step-by-Step Solution**

    1. Calculate the energy contribution from the carbohydrates.

$$4 \times 12.5 = 50$$

    Carbohydrates contribute 50 Cal to the energy content of the glass of milk.

    2. Calculate the energy contribution from the fat.

$$9 \times 1.4 = 12.6$$

    Fats contribute 12.6 Cal to the energy content of the glass of milk.

    3. Calculate the energy contribution from the protein.

$$4 \times 4.2 = 16.8$$

    Proteins contribute 16.8 Cal to the energy content of the glass of milk.

    4. Calculate the energy content of the glass of milk.

$$50 + 12.6 + 16.8 = 79.4$$

    The glass of milk contains 79.4 Cal of energy.

An Algebraic Solution

E energy content of the glass of milk (Cal)
$E = 4 \times 12.5 + 9 \times 1.4 + 4 \times 4.2$

$E = 50 + 12.6 + 16.8$
$E = 79.4$

The glass of milk contains 79.4 Cal of energy.

o. **A Step-by-Step Solution**

    1. Calculate the energy contribution from the carbohydrates.

$$4 \times 16 = 64$$

    Carbohydrates contribute 64 Cal to the energy content of the cup of yogurt.

    2. Calculate the energy contribution from the fat.

$$9 \times 4.2 = 37.8$$

    Fats contribute 37.8 Cal to the energy content of the cup of yogurt.

3. Calculate the energy content of the cup of yogurt.

$$64 + 37.8 = 101.8$$

The cup of yogurt contains 101.8 Cal of energy.

An Algebraic Solution

E energy content of the cup of yogurt (Cal)
$E = 4 \times 16 + 9 \times 4.2$

$E = 64 + 37.8$
$E = 101.8$

The cup of yogurt contains 101.8 Cal of energy.

p. **A Step-by-Step Solution**

1. Calculate the energy gain from the slices of bread.

$$37 \times 3 = 111$$

Marco had a gain of 111 Cal of energy from the slices of bread.

2. Calculate the energy gain from the jam.

$$52 \times 2 = 104$$

Marco had a gain of 104 Cal of energy from the jam.

3. Calculate the energy loss from resting.

$$80 \times 2 = 160$$

Marco had a loss of 160 Cal of energy from resting.

4. Calculate the energy loss from exercising.

$$320 \times 0.5 = 160$$

Marco had a loss of 160 Cal of energy from exercising.

5. Calculate Marco's net intake of energy since breakfast.

$$111 + 104 + 20 - 160 - 160 = -85$$

Since breakfast, Marco had a net intake of -85 Cal of energy. Alternatively, one may say that since breakfast, Marco had a net loss of 85 Cal of energy.

An Algebraic Solution

E Marco's net intake of energy since breakfast (Cal)
$E = 37 \times 3 + 52 \times 2 + 20 - 80 \times 2 - 320 \times 0.5$

$E = 111 + 104 + 20 - 160 - 160$
$E = -85$

Since breakfast, Marco had a net intake of -85 Cal of energy. Alternatively, one may say that since breakfast, Marco had a net loss of 85 Cal of energy.

q. **A Step-by-Step Solution**

1. Calculate the energy contribution from the cheese.

   $80 \times 3 = 240$

   The slices of cheese contain 240 Cal of energy.

2. Calculate the energy contribution from the slices of toast.

   $40 \times 2 = 80$

   The slices of toast contain 80 Cal of energy.

3. Calculate the energy contribution from the coffee.

   $90 \times 2 = 180$

   The coffee contains 180 Cal of energy.

4. Calculate the energy content of the breakfast deal.

   $240 + 80 + 180 + 120 = 620$

   The breakfast deal contains 620 Cal of energy.

An Algebraic Solution

E energy content of the breakfast deal (Cal)
$E = 80 \times 3 + 40 \times 2 + 90 \times 2 + 120$

$E = 240 + 80 + 180 + 120$
$E = 620$

The breakfast deal contains 620 Cal of energy.

2. The following are possible word problems whose solutions can be modelled as the given equations.

   a. It costs \$4 to mail a package and \$15 to mail a box. Calculate the cost of mailing 2 packages and 3 boxes.

      c cost of mailing the packages and boxes (\$)
      $c = 4 \times 2 + 15 \times 3$

b. A garden is designed as a circle with 4 surrounding squares. The circle has an area of 12.99 m^2 and each square has an area of 2.5 m^2. Calculate the total area of the garden.

A area of the garden (m^2)
$A = 12.99 + 2.5 \times 4$

c. 1020 apples arrived at the plant at 8 am. The machines have been cleaning and tagging apples at a rate of 430 apples/h for the past 2 h. Another set of 500 apples arrived at the plant at 10 am. How many apples are waiting to be cleaned and tagged?

n number of apples waiting to be cleaned and tagged (1)
$n = 1020 - 430 \times 2 + 500$

d. You have a 3.1 m beam with a linear density of 12.43 g/m, a 4.2 m beam with a linear density of 9.2 g/m and a 2.5 m beam with a linear density of 3.8 g/m. What is the total mass of the three beams?

m mass of the three beams (g)
$m = 12.43 \times 3.1 + 9.2 \times 4.2 + 3.8 \times 2.5$

e. This morning the temperature was 20.5 °C. The temperature dropped by 2.5 °C and then rose by 4.2 °C. What is the temperature now?

t current value of temperature (°C)
$t = 20.5 - 2.5 + 4.2$

f. The house has 5 floors. Each floor has 4 units. Each unit has 3 windows. How many windows does this house have?

n number of windows in this house (1)
$n = 3 \times 4 \times 5$

g. Calculate the volume of a container in the shape of a rectangular solid with a length of 2 m, a width of 3.1 m and a height of 5.99 m.

V volume of the container (m^3)
$V = 5.99 \times 3.1 \times 2$

Exercise Set 2.2

1. A Step-by-Step Solution

1. Calculate total amount of money spent.

    a. Calculate the amount of money spent on the pens.

    $$2 \times 3 = 6$$

    I spent $6 on the pens.

    b. Calculate total amount of money spent.

    $$16.99 + 6 = 22.99$$

    I spent $22.99 in total.

2. Calculate amount of money left.

    $$58 - 22.99 = 35.01$$

    I have $35.01 left.

An Algebraic Solution

a amount of money left after purchases ($)
$a = 58 - (16.99 + 2 \times 3)$

$a = 58 - (16.99 + 6)$
$a = 58 - 22.99$
$a = 35.01$

I have $35.01 left.

2. A Step-by-Step Solution

1. Calculate total amount of payments.

    a. Calculate the total amount of the three payments.

    $$52.50 \times 3 = 157.50$$

    The three payments add up to $157.50.

    b. Calculate the total amount of the two payments.

    $$30.99 \times 2 = 61.98$$

    The two payments add up to $61.98.

    c. Calculate total amount of payments.

    $$157.50 + 61.98 = 219.48$$

    The payments add up to $219.48.

2. Calculate amount of money left after payments.

$$420 - 219.48 = 200.52$$

After the payments, I have $200.52 left.

An Algebraic Solution

a amount left after payments ($)
$a = 420 - (52.50 \times 3 + 30.99 \times 2)$

$a = 420 - (157.50 + 61.98)$
$a = 420 - 219.48$
$a = 200.52$

After the payments, I have $200.52 left.

3. **A Step-by-Step Solution**

1. Calculate the volume of medication used.
   a. Calculate the volume of medication given to the eight patients.

   $$20.5 \times 8 = 164$$

   164 mL of medication was given to the eight patients.
   b. Calculate the volume of medication given to the twelve patients.

   $$12.5 \times 12 = 150$$

   150 mL of medication was given to the twelve patients.
   c. Calculate the volume of medication given to the eighteen patients.

   $$3.5 \times 18 = 63$$

   63 mL of medication was given to the eighteen patients.
   d. Calculate the volume of medication used.

   $$164 + 150 + 63 = 377$$

   In total, 377 mL of medication was used.
2. Calculate the volume of medication left.

   $$657 - 377 = 280$$

   280 mL of medication is left.

An Algebraic Solution

V volume of medication left (mL)
$V = 657 - (20.5 \times 8 + 12.5 \times 12 + 3.5 \times 18)$

$V = 657 - (164 + 150 + 63)$
$V = 657 - 377$
$V = 280$

280 mL of medication is left.

4. **A Step-by-Step Solution**

   1. Calculate total volume of salt solution used.
      a. Calculate volume of salt solution used this morning.

      $$1.25 \times 1.5 = 1.875$$

      1.875 L of the salt solution was used this morning.
      b. Calculate volume of salt solution used this afternoon.

      $$0.57 \times 2.5 = 1.425$$

      1.425 L of the salt solution was used this afternoon.
      c. Calculate total volume of salt solution used.

      $$1.875 + 1.425 = 3.3$$

      In total, 3.3 L of salt solution was used.
   2. Calculate volume of salt solution left.

      $$8.1 - 3.3 = 4.8$$

      4.8 L of salt solution is left.

An Algebraic Solution

V volume of salt solution left (L)
$V = 8.1 - (1.25 \times 1.5 + 0.57 \times 2.5)$

$V = 8.1 - (1.875 + 1.425)$
$V = 8.1 - 3.3$
$V = 4.8$

4.8 L of salt solution is left.

5. **A Step-by-Step Solution**

    1. Calculate the mass of a molecule of SO_2.

        a. Calculate the mass of the O atoms in a molecule of SO_2.

$$16.00 \times 2 = 32$$

        The O atoms have a mass of 32 amu.

        b. Calculate the mass of a molecule of SO_2.

$$32.07 + 32 = 64.07$$

        A molecule of SO_2 has a mass of 64.07 amu.

    2. Calculate the mass of the molecules of SO_2.

$$64.07 \times 42 = 2690.94$$

    The SO_2 molecules have a mass of 2690.94 amu.

An Algebraic Solution

m mass of the molecules of SO_2 (amu)
$m = 42\,(32.07 + 16.00 \times 2)$

$m = 42\,(32.07 + 32)$
$m = 42 \times 64.07$
$m = 2690.94$

The SO_2 molecules have a mass of 2690.94 amu.

6. **A Step-by-Step Solution**

    1. Calculate the mass of 1 mol of SO_2.

        a. Calculate the mass of the O atoms in 1 mol of SO_2.

$$16.00 \times 2 = 32$$

        The O atoms have a mass of 32 g.

        b. Calculate the mass of 1 mol of SO_2.

$$32.07 + 32 = 64.07$$

        The mass of 1 mol of SO_2 is 64.07 g.

    2. Calculate the mass of the molecules of SO_2.

$$64.07 \times 42 = 2690.94$$

    The SO_2 molecules have a mass of 2690.94 g.

An Algebraic Solution

m mass of the molecules of SO_2 (g)
$m = 42\,(32.07 + 16.00 \times 2)$

$m = 42\,(32.07 + 32)$
$m = 42 \times 64.07$
$m = 2690.4$

The SO_2 molecules have a mass of 2690.4 g.

7. **A Step-by-Step Solution**

1. Calculate the mass of OH groups in a molecule of $Ca(OH)_2$.
   a. Calculate the mass of one OH group.

   $$16.00 + 1.008 = 17.008$$

   An OH group has a mass of 17.008 amu.
   b. Calculate the mass of OH groups in a molecule of $Ca(OH)_2$.

   $$17.008 \times 2 = 34.016$$

   The OH groups have a mass of 34.016 amu.
2. Calculate the mass of a molecule of $Ca(OH)_2$.

   $$40.08 + 34.016 = 74.096$$

   A molecule of $Ca(OH)_2$ has a mass of 74.096 amu.

An Algebraic Solution

m mass of a molecule of $Ca(OH)_2$ (amu)
$m = 40.08 + 2\,(16.00 + 1.008)$

$m = 40.08 + 2 \times 17.008$
$m = 40.08 + 34.016$
$m = 74.096$

A molecule of $Ca(OH)_2$ has a mass of 74.096 amu.

8. **A Step-by-Step Solution**

1. Calculate the mass of Li atoms in 1 mol of $Li_2(CO_3)_4$.

   $$6.941 \times 2 = 13.882$$

   The Li atoms have a mass of 13.882 g.

2. Calculate the mass of CO_3 groups in 1 mol of $Li_2(CO_3)_4$.
   a. Calculate the mass of 1 mol of CO_3.
      i. Calculate the mass of O atoms in 1 mol of CO_3.

$$16.00 \times 3 = 48$$

The O atoms have a mass of 48 g.
      ii. Calculate the mass of 1 mol of CO_3.

$$12.01 + 48 = 60.01$$

The mass of 1 mol of CO_3 is 60.01 g.
   b. Calculate the mass of CO_3 groups in 1 mol of $Li_2(CO_3)_4$.

$$60.01 \times 4 = 240.04$$

The CO_3 groups have a mass of 240.04 g.
3. Calculate the mass of 1 mol of $Li_2(CO_3)_4$.

$$13.882 + 240.04 = 253.922$$

The mass of 1 mol of $Li_2(CO_3)_4$ is 253.922 g.

An Algebraic Solution

m mass of 1 mol of $Li_2(CO_3)_4$ (g)
$m = 6.941 \times 2 + 4(12.01 + 16.00 \times 3)$
$m = 13.882 + 4(12.01 + 48)$
$m = 13.882 + 4 \times 60.01$
$m = 13.882 + 240.04$
$m = 253.922$

The mass of 1 mol of $Li_2(CO_3)_4$ is 253.922 g.

9. **A Step-by-Step Solution**
   1. Calculate the mass of 1 mol of $Na_4(PO_4)_3$.
      a. Calculate the mass of Na atoms in 1 mol of $Na_4(PO_4)_3$.

$$22.99 \times 4 = 91.96$$

The Na atoms have a mass of 91.96 g.
      b. Calculate the mass of PO_4 groups in 1 mol of $Na_4(PO_4)_3$.
         i. Calculate the mass of 1 mol of PO_4.
            α. Calculate the mass of O atoms in 1 mol of PO_4.

$$16.00 \times 4 = 64$$

The O atoms have a mass of 64 g.
            β. Calculate the mass of 1 mol of PO_4.

$$30.97 + 64 = 94.97$$

The mass of 1 mol of PO_4 is 94.97 g.

ii. Calculate the mass of PO_4 groups in 1 mol of $Na_4(PO_4)_3$.

$$94.97 \times 3 = 284.91$$

The PO_4 groups have a mass of 284.91 g.

c. Calculate the mass of 1 mol of $Na_4(PO_4)_3$.

$$91.96 + 284.91 = 376.87$$

The mass of 1 mol of $Na_4(PO_4)_3$ is 376.87 g.

2. Calculate the mass of 4.8 mol of $Na_4(PO_4)_3$.

$$376.87 \times 4 = 1808.976$$

The mass of 4.8 mol of $Na_4(PO_4)_3$ is 1808.976 g.

An Algebraic Solution

m mass of 1 mol of $Na_4(PO_4)_3$ (g)

$m = 4.8 \left[22.99 \times 4 + 3\,(30.96 + 16.00 \times 4) \right]$

$m = 4.8 \left[91.96 + 3\,(30.96 + 64) \right]$

$m = 4.8\,(91.96 + 3 \times 94.96)$

$m = 4.8\,(91.96 + 284.91)$

$m = 4.8 \times 376.87$

$m = 1808.976$

The mass of 4.8 mol of $Na_4(PO_4)_3$ is 1808.976 g.

10. **A Step-by-Step Solution**

1. Calculate the mass of a molecule of $(NH_4)_3PO_4$.

a. Calculate the mass of NH_4 groups in a molecule of $(NH_4)_3PO_4$.

i. Calculate the mass of one group of NH_4.

α. Calculate the mass of H atoms in one group of NH_4.

$$1.008 \times 4 = 4.032$$

The H atoms have a mass of 4.032 amu.

β. Calculate the mass of one group of NH_4.

$$14.01 + 4.032 = 18.042$$

The mass of one group of NH_4 is 18.042 amu.

ii. Calculate the mass of NH_4 groups in a molecule of $(NH_4)_3PO_4$.

$$18.042 \times 3 = 54.126$$

The NH_4 groups have a mass of 54.126 amu.

b. Calculate the mass of O atoms in a molecule of $(NH_4)_3PO_4$.

$$16 \times 4 = 64$$

The O atoms have a mass of 64 amu.

c. Calculate the mass of a molecule of $(NH_4)_3PO_4$.

$$54.126 + 30.97 + 64 = 149.096$$

The mass of a molecule of $(NH_4)PO_4$ is 149.096 amu.

2. Calculate the mass of 158 molecules of $(NH_4)_3PO_4$.

$$149.096 \times 158 = 23\,557.168$$

The mass of 158 molecules of $(NH_4)_3PO_4$ is 23 557.168 amu.

An Algebraic Solution

m mass of 158 molecules of $(NH_4)_3PO_4$ (amu)

$m = 158\left[3\,(14.01 + 1.008 \times 4) + 30.97 + 16.00 \times 4\right]$

$m = 158\left[3\,(14.01 + 4.032) + 30.97 + 64\right]$

$m = 158\,(3 \times 18.042 + 30.97 + 64)$

$m = 158\,(54.126 + 30.97 + 64)$

$m = 158 \times 149.096$

$m = 23\,557.168$

The mass of 158 molecules of $(NH_4)_3PO_4$ is 23 557.168 amu.

Exercise Set 2.3

1. **A Step-by-Step Solution**

1. Calculate the sum of the temperatures.

$$12 + 15 + 18 + 14 + 20 = 79$$

The sum of the temperatures is 79 °C.

2. Calculate the average temperature.

$$79 \div 5 = 15.8$$

The average temperature over the past five days was 15.8 °C.

An Algebraic Solution

$\bar{t}$ average temperature over the past five days (°C)

$$\bar{t} = \frac{12 + 15 + 18 + 14 + 20}{5}$$

$$\bar{t} = \frac{79}{5}$$

$$\bar{t} = 15.8$$

The average temperature over the past five days was 15.8 °C.

2. **A Step-by-Step Solution**

   1. Calculate the sum of the expenses.

$$2100 + 1600 + 1950 = 5650$$

The sum of the expenses is \$5650.

   2. Calculate the average expense.

$$5650 \div 3 = 1883.333 \cdots$$

My expenses over the past three months average \$1883.33/month.

An Algebraic Solution

$\bar{a}$ average expense over the past three months (\$/month)

$$\bar{a} = \frac{2100 + 1600 + 1950}{3}$$

$$\bar{a} = \frac{5650}{3}$$

$$\bar{a} = 1883.333 \cdots$$

My expenses over the past three months average \$1883.33/month.

3. **A Step-by-Step Solution**

   1. Calculate the cost of all computer systems.

     a. Calculate the cost of Type A computer systems.

$$1200 \times 8 = 9600$$

The cost of Type A computer systems is \$9600.

     b. Calculate the cost of Type C computer systems.

$$700 \times 3 = 2100$$

The cost of Type C computer systems is \$2100.

c. Calculate the cost of all computer systems.

$$9600 + 820 + 2100 = 12\,520$$

The computer systems cost $12\,520.

2. Calculate the total number of computer systems.

$$8 + 1 + 3 = 12$$

In total, there are 12 computer systems.

3. Calculate the average cost of a computer system.

$$12\,520 \div 12 = 1043.333\cdots$$

The average cost of a computer system is $1043.33/system.

An Algebraic Solution

$\bar{c}$ average cost of a computer system ($/system)

$$\bar{c} = \frac{1200 \times 8 + 820 + 700 \times 3}{8 + 1 + 3}$$

$$\bar{c} = \frac{9600 + 820 + 2100}{12}$$

$$\bar{c} = \frac{12\,520}{12}$$

$$\bar{c} = 1043.333\cdots$$

The average cost of a computer system is $1043.33/system.

4. A Step-by-Step Solution

1. Calculate total income.
   a. Calculate income from first job.

   $$17 \times 20 = 340$$

   I made $340 from my first job.
   b. Calculate income from second job.

   $$21 \times 12 = 252$$

   I made $252 from my second job.
   c. Calculate total income.

   $$340 + 252 = 592$$

   I made $592 in total.
2. Calculate total time worked.

   $$20 + 12 = 32$$

In total I worked 32 h.

3. Calculate the average rate of pay per hour.

$$592 \div 32 = 18.50$$

On average I made $18.50/h.

An Algebraic Solution

$\bar{r}$ average rate of income ($/h)

$$\bar{r} = \frac{17 \times 20 + 21 \times 12}{20 + 12}$$

$$\bar{r} = \frac{340 + 252}{32}$$

$$\bar{r} = \frac{592}{32}$$

$$\bar{r} = 18.50$$

On average I made $18.50/h.

5. A Step-by-Step Solution

1. Calculate total income.
   a. Calculate income from first job.

   $$14 \times 15 = 210$$

   I made $210 from my first job.
   b. Calculate income from second job.

   $$16 \times 10 = 160$$

   I made $160 from my second job.
   c. Calculate income from third job.

   $$18 \times 2 = 36$$

   I made $36 from my third job.
   d. Calculate total income.

   $$210 + 160 + 36 = 406$$

   In total I made $406.
2. Calculate time worked.

   $$15 + 10 + 2 = 27$$

   In total I worked 27 hours.
3. Calculate the average rate of pay.

   $$406 \div 27 = 15.04$$

   On average I made $15.04/h.

An Algebraic Solution

$\bar{r}$ average rate of income (\$/h)

$$\bar{r} = \frac{14 \times 15 + 16 \times 10 + 18 \times 2}{15 + 10 + 2}$$

$$\bar{r} = \frac{210 + 160 + 36}{27}$$

$$\bar{r} = \frac{406}{27}$$

$$\bar{r} = 15.04$$

On average I made \$15.04/h.

6. **A Step-by-Step Solution**[2]

    1. Calculate total training time per week.

       a. Calculate total training time per week for workers in Section A.

$$3 \times 120 = 360$$

Total training time per week for workers in Section A is 360 h/week.

       b. Calculate total training time per week for workers in Section B.

$$5 \times 250 = 1250$$

Total training time per week for workers in Section B is 1250 h/week.

       c. Calculate total training time per week for workers in Section C.

$$8 \times 320 = 2560$$

Total training time per week for workers in Section C is 2560 h/week.

       d. Calculate training time per week for all workers.

$$360 + 1250 + 2560 = 4170$$

Total training time per week for all the workers is 4170 h/week.

    2. Calculate total number of workers.

$$120 + 250 + 320 = 690$$

In total there are 690 workers.

---

[2]The unit used in the solution below, i.e., h/week/worker, is often written as h/(week·worker). We let the reader apply the simplification rules for simplifying complex fractions to the former expression to derive the latter. Note that the former presentation is more descriptive while the latter expression is more efficient.

3. Calculate average training time per week per worker.

$$4170 \div 690 = 6.043\cdots$$

The average training time per week per worker is
6 h/week/worker.

An Algebraic Solution

$\bar{t}$ average training time per week per worker (h/week/worker)

$$\bar{t} = \frac{3 \times 120 + 5 \times 250 + 8 \times 320}{120 + 250 + 320}$$

$$\bar{t} = \frac{360 + 1250 + 2560}{690}$$

$$\bar{t} = \frac{4170}{690}$$

$$\bar{t} = 6.043$$

The average training time per week per worker is
6 h/week/worker.

7. **A Step-by-Step Solution**

1. Calculate total annual salary of all workers.
   a. Calculate total annual salary of workers in Section A.

   $$64\,000 \times 56 = 3\,584\,000$$

   The total annual salary of workers in Section A is
   \$3 584 000/year.
   b. Calculate total annual salary of workers in Section B.

   $$45\,000 \times 120 = 5\,400\,000$$

   The total annual salary of workers in Section B is
   \$5 400 000/year.
   c. Calculate total annual salary of workers in Section C.

   $$52\,000 \times 98 = 5\,096\,000$$

   The total annual salary of workers in Section C is
   \$5 096 000/year.
   d. Calculate total annual salary of all workers.

   $$3\,584\,000 + 5\,400\,000 + 5\,096\,000 = 14\,080\,000$$

   The total annual salary of all workers is \$14 080 000/year.
2. Calculate total number of workers.

   $$56 + 120 + 98 = 274$$

   In total, there are 274 workers at this company.

3. Calculate average annual salary of a worker.

$$14\,080\,000 \div 274 = 51\,386.86$$

The average annual salary of a worker at this company is $51\,386.86/year/worker.

An Algebraic Solution

$\overline{a}$ average salary of a worker at this company ($/year/worker)

$$\overline{a} = \frac{64\,000 \times 56 + 45\,000 \times 120 + 52\,000 \times 98}{56 + 120 + 98}$$

$$\overline{a} = \frac{3\,584\,000 + 5\,400\,000 + 5\,096\,000}{274}$$

$$\overline{a} = \frac{14\,080\,000}{274}$$

$$\overline{a} = 51\,386.86$$

The average salary of a worker at this company is $51\,386.86/year/worker.

8. **A Step-by-Step Solution**

  1. Calculate total grade points earned.

    a. Calculate grade points earned from the MATH101 course.

$$3.33 \times 3 = 9.99$$

Jorge earned 9.99 grade points from the MATH101 course.

    b. Calculate grade points earned from the BIOL101 course.

$$1.67 \times 5 = 8.35$$

Jorge earned 8.35 grade points from the BIOL101 course.

    c. Calculate grade points earned from the CHEM101 course.

$$1.00 \times 4 = 4$$

Jorge earned 4 grade points from the CHEM101 course.

    d. Calculate total grade points earned.

$$9.99 + 8.35 + 4 = 22.34$$

In total, Jorge earned 22.34 grade points.

  2. Calculate total credit value.

$$3 + 5 + 4 = 12$$

The total credit value is 12.

3. Calculate Jorge's term GPA this term.

$$22.34 \div 12 = 1.86$$

Jorge's term GPA this term is 1.86.

An Algebraic Solution

g Jorge's term GPA this term (1)

$$g = \frac{3.33 \times 3 + 1.67 \times 5 + 1.00 \times 4}{3 + 5 + 4}$$

$$g = \frac{9.99 + 8.35 + 4}{12}$$

$$g = \frac{22.34}{12}$$

$$g = 1.86$$

Jorge's term GPA this term is 1.86.

9. **A Step-by-Step Solution**

1. Calculate total grade points .
   a. Calculate grade points earned from the PHOT1148 course.

   $$4.00 \times 3 = 12$$

   I earned 12 grade points from the PHOT1148 course.
   b. Calculate grade points earned from the PSY1040 course.

   $$2.67 \times 2 = 5.34$$

   I earned 5.34 grade points from the PSY1040 course.
   c. Calculate grade points earned from the HUM1014 course.

   $$2.33 \times 4 = 9.32$$

   I earned 9.32 grade points from the HUM1014 course.
   d. Calculate grade points earned from the COMM1017 course.

   $$3.67 \times 3 = 11.01$$

   I earned 11.01 grade points from the COMM1017 course.
   e. Calculate total grade points earned.

   $$12 + 5.34 + 9.32 + 11.01 = 37.67$$

   In total, I earned 37.67 grade points.
2. Calculate total credit value.

   $$3 + 2 + 4 + 3 = 12$$

   The total credit value is 12.

3. Calculate my term GPA last term.

$$37.67 \div 12 = 3.1391 \cdots$$

Last term, I had a term GPA of 3.14.

An Algebraic Solution

g My term GPA last term (1)

$$g = \frac{4.00 \times 3 + 2.67 \times 2 + 2.33 \times 4 + 3.67 \times 3}{3 + 2 + 4 + 3}$$

$$g = \frac{12.00 + 5.34 + 9.32 + 11.01}{12}$$

$$g = \frac{37.67}{12}$$

$$g = 3.1391 \cdots$$

Last term, I had a term GPA of 3.14.

10. **A Step-by-Step Solution**

1. Calculate the mass of sulfur atoms in the sample.
   a. Calculate the mass of ^{32}S isotopes in the sample.

   $$32 \times 94.93 = 3037.76$$

   The ^{32}S isotopes of sulfur have a mass of 3037.76 amu.
   b. Calculate the mass of ^{33}S isotopes in the sample.

   $$33 \times 0.76 = 25.08$$

   The ^{33}S isotopes of sulfur have a mass of 25.08 amu.
   c. Calculate the mass of ^{34}S isotopes in the sample.

   $$34 \times 4.29 = 145.86$$

   The ^{34}S isotopes of sulfur have a mass of 145.86 amu.
   d. Calculate the mass of ^{36}S isotopes in the sample.

   $$36 \times 0.02 = 0.72$$

   The ^{36}S isotopes of sulfur have a mass of 0.72 amu.
   e. Calculate the mass of sulfur atoms in the sample.

   $$3037.76 + 25.08 + 145.86 + 0.72 = 3209.42$$

   The sulfur atoms have a mass of 3209.42 amu.
2. Calculate total number of sulfur atoms in the sample.

   $$94.93 + 0.76 + 4.29 + 0.02 = 100$$

   There are 100 sulfur atoms in the sample.

3. Calculate the atomic mass of sulfur.

$$3209.42 \div 100 \ = \ 32.0942$$

The atomic mass of sulfur is 32.09 amu/atom.

An Algebraic Solution

m_a atomic mass of sulfur (amu/atom)

$$m_a \ = \ \frac{32 \times 94.93 \ + \ 33 \times 0.76 \ + \ 34 \times 4.29 \ + \ 36 \times 0.02}{94.93 \ + \ 0.76 \ + \ 4.29 \ + \ 0.02}$$

$$m_a \ = \ \frac{3037.76 \ + \ 25.08 \ + \ 145.86 \ + \ 0.72}{100}$$

$$m_a \ = \ \frac{3209.42}{100}$$

$$m_a \ = \ 32.0942$$

The atomic mass of sulfur is 32.09 amu/atom.

11. A Step-by-Step Solution

1. Calculate the mass of silver atoms in the sample.
   a. Calculate the mass of ^{107}Ag isotopes in the sample.

   $$107 \times 51.839 \ = \ 5546.773$$

   The ^{107}Ag isotopes of silver have a mass of 5546.773 amu.
   b. Calculate the mass of ^{109}Ag isotopes in the sample.

   $$109 \times 48.161 \ = \ 5249.549$$

   The ^{109}Ag isotopes of silver have a mass of 5249.549 amu.
   c. Calculate the mass of silver atoms in the sample.

   $$5546.773 \ + \ 5249.549 \ = \ 10\,796.322$$

   The silver atoms have a mass of 10 796.322 amu.
2. Calculate total number of silver atoms in the sample.

   $$51.839 \ + \ 48.161 \ = \ 100$$

   There are 100 silver atoms in the sample.
3. Calculate the atomic mass of silver.

   $$10\,796.322 \div 100 \ = \ 107.963\,22$$

   The atomic mass of silver is 107.963 amu/atom.

An Algebraic Solution

m_a atomic mass of silver (amu/atom)

$$m_a = \frac{107 \times 51.839 \,+\, 109 \times 48.161}{51.839 \,+\, 48.161}$$

$$m_a = \frac{5546.773 \,+\, 5249.549}{100}$$

$$m_a = \frac{10796.322}{100}$$

$$m_a = 107.963\,22$$

The atomic mass of silver is 107.963 amu/atom.

12. **A Step-by-Step Solution**

   1. Calculate the mass of oxygen atoms in the sample.

      a. Calculate the mass of ^{16}O isotopes in the sample.

$$16 \times 64.2 = 1027.2$$

      The ^{16}O isotopes of oxygen have a mass of 1027.2 amu.

      b. Calculate the mass of ^{17}O isotopes in the sample.

$$17 \times 35.8 = 608.6$$

      The ^{17}O isotopes of oxygen have a mass of 608.6 amu.

      c. Calculate the mass of oxygen atoms in the sample.

$$1027.2 + 608.6 = 1635.8$$

      The oxygen atoms have a mass of 1635.8 amu.

   2. Calculate total number of oxygen atoms in the sample.

$$64.2 + 35.8 = 100$$

   There are 100 oxygen atoms in the sample.

   3. Calculate the average mass of an oxygen atom from the sample.

$$1635.8 \div 100 = 16.358$$

   The average mass of an oxygen atom in the sample is 16.358 amu/atom.

An Algebraic Solution

$\overline{m}$ average mass of an oxygen atom in the sample (amu/atom)

$$\overline{m} = \frac{16 \times 64.2 \ + \ 17 \times 35.8}{64.2 \ + \ 35.8}$$

$$\overline{m} = \frac{1027.2 \ + \ 608.6}{100}$$

$$\overline{m} = \frac{1635.8}{100}$$

$$\overline{m} = 16.358$$

The average mass of an oxygen atom in the sample is 16.358 amu/atom.

13. **A Step-by-Step Solution**

1. Calculate the mass of nitrogen atoms in the sample.

    a. Calculate the mass of ^{14}N isotopes in the sample.

    $$14 \times 82 = 1148$$

    The ^{14}N isotopes of nitrogen have a mass of 1148 amu.

    b. Calculate the mass of ^{16}N isotopes in the sample.

    $$16 \times 13 = 208$$

    The ^{16}N isotopes of nitrogen have a mass of 208 amu.

    c. Calculate the mass of ^{17}N isotopes in the sample.

    $$17 \times 5 = 85$$

    The ^{17}N isotopes of nitrogen have a mass of 85 amu.

    d. Calculate the mass of nitrogen atoms in the sample.

    $$1148 \ + \ 208 \ + \ 85 = 1441$$

    The nitrogen atoms have a mass of 1441 amu.

2. Calculate total number of nitrogen atoms in the sample.

    $$82 \ + \ 13 \ + \ 5 = 100$$

    There are 100 nitrogen atoms in the sample.

3. Calculate the average mass of a nitrogen atom from the sample.

    $$1441 \div 100 = 14.41$$

    The average mass of a nitrogen atom in the sample is 14.41 amu/atom.

An Algebraic Solution

$\overline{m}$ average mass of a nitrogen atom in the sample (amu/atom)

$$\overline{m} = \frac{14 \times 82 + 16 \times 13 + 17 \times 5}{82 + 13 + 5}$$

$$\overline{m} = \frac{1148 + 208 + 85}{100}$$

$$\overline{m} = \frac{1441}{100}$$

$$\overline{m} = 14.41$$

The average mass of a nitrogen atom in the sample is 14.41 amu/atom.

Exercise Set 3.1

1. a. The expression on the right side of the equation that models the problem, i.e., $1.99 \times 3 + 12.80$, is a superposition of the two terms 1.99×3 and 12.80.

   b. The expression on the right side of the equation that models the problem, i.e., 16.00×3, is a superposition of the one term 16.00×3.

   c. The expression on the right side of the equation that models the problem, i.e., $1.008 \times 2 + 16.00$, is a superposition of the two terms 1.008×2 and 16.00.

   d. The expression on the right side of the equation that models the problem, i.e., $32.80 - (1.30 \times 2 + 3.50 \times 4)$, is a superposition of the two terms 32.80 and $(1.30 \times 2 + 3.50 \times 4)$. The second term contains the single factor $(1.30 \times 2 + 3.50 \times 4)$ which is a superposition of the two terms 1.30×2 and 3.50×4.

   e. The expression on the right side of the equation that models the problem, i.e., $4 \times 21.6 + 9 \times 12 + 4 \times 10.2$, is a superposition of the three terms 4×21.6, 9×12 and 4×10.2.

   f. The expression on the right side of the equation that models the problem, i.e., $32.4 - (1.25 \times 2.5 + 3.25 \times 1.5)$, is a superposition of the two terms 32.4 and $(1.25 \times 2.5 + 3.25 \times 1.5)$. The second term contains the single factor $(1.25 \times 2.5 + 3.25 \times 1.5)$ which is a superposition of the two terms 1.25×2.5 and 3.25×1.5.

   g. The expression on the right side of the equation that models the problem, i.e., $4.5\,(12.01 \times 2 + 1.008 \times 4)$, is a superposition of the one term $4.5\,(12.01 \times 2 + 1.008 \times 4)$. The single term $4.5\,(12.01 \times 2 + 1.008 \times 4)$ breaks into the two factors 4.5 and $(12.01 \times 2 + 1.008 \times 4)$ the second of which is a superposition of the two terms 12.01×2 and 1.008×4.

   h. The expression on the right side of the equation that models the problem, i.e., $120 \times 1.5 + 90 \times 2.5$, is a superposition of the two terms 120×1.5 and 90×2.5.

   i. The expression on the right side of the equation that models the problem, i.e., $1.82 - 0.35 \times 3$, is a superposition of the two terms 1.82 and 0.35×3.

j. The expression on the right side of the equation that models the problem, i.e., $\frac{20.5 \times 2 + 8.5 \times 3}{2+3}$, is a superposition of the one term $\frac{20.5 \times 2 + 8.5 \times 3}{2+3}$. The numerator of this expression, i.e., $20.5 \times 2 + 8.5 \times 3$, is a superposition of the two terms 20.5×2 and 8.5×3 and the denominator of this expression, i.e., $2 + 3$, is a superposition of the two terms 2 and 3.

k. The expression on the right side of the equation that models the problem, i.e., $16 \times 0.973 + 17 \times 0.026 + 18 \times 0.001$, is a superposition of the three terms 16×0.973, 17×0.026 and 18×0.001.

l. The expression on the right side of the equation that models the problem, i.e., $\frac{110 \times 1.5 + 85 \times 3}{1.5+3}$, is a superposition of the one term $\frac{110 \times 1.5 + 85 \times 3}{1.5+3}$. The numerator of this expression, i.e., $110 \times 1.5 + 85 \times 3$, is a superposition of the two terms 110×1.5 and 85×3. The denominator of this expression, i.e., $1.5 + 3$, is a superposition of the two terms 1.5 and 3.

Exercise Set 3.2

1. a. The expression on the right side of the equation that models the problem, i.e., $1.99 \times 3 + 12.80$, is a superposition of the two terms 1.99×3 and 12.80.

b. The expression on the right side of the equation that models the problem, i.e., 16.00×3, is a superposition of the one term 16.00×3.

c. The expression on the right side of the equation that models the problem, i.e., $1.008 \times 2 + 16.00$, is a superposition of the two terms 1.008×2 and 16.00.

d. The expression on the right side of the equation that models the problem, i.e., $32.80 - (1.30 \times 2 + 3.50 \times 4)$, is a superposition of the two terms 32.80 and $(1.30 \times 2 + 3.50 \times 4)$.

e. The expression on the right side of the equation that models the problem, i.e., $4 \times 21.6 + 9 \times 12 + 4 \times 10.2$, is a superposition of the three terms 4×21.6, 9×12 and 4×10.2.

f. The expression on the right side of the equation that models the problem, i.e., $32.4 - (1.25 \times 2.5 + 3.25 \times 1.5)$, is a superposition of the two terms 32.4 and $(1.25 \times 2.5 + 3.25 \times 1.5)$.

g. The expression on the right side of the equation that models the problem, i.e., $4.5\,(12.01 \times 2 + 1.008 \times 4)$, is a superposition of the one term $4.5\,(12.01 \times 2 + 1.008 \times 4)$.

h. The expression on the right side of the equation that models the problem, i.e., $120 \times 1.5 + 90 \times 2.5$, is a superposition of the two terms 120×1.5 and 90×2.5.

i. The expression on the right side of the equation that models the problem, i.e., $1.82 - 0.35 \times 3$, is a superposition of the two terms 1.82 and 0.35×3.

j. The expression on the right side of the equation that models the problem, i.e., $\frac{20.5 \times 2 + 8.5 \times 3}{2+3}$, is a superposition of the one term $\frac{20.5 \times 2 + 8.5 \times 3}{2+3}$.

k. The expression on the right side of the equation that models the problem, i.e., $16 \times 0.973 + 17 \times 0.026 + 18 \times 0.001$, is a superposition of the three terms 16×0.973, 17×0.026 and 18×0.001.

1. The expression on the right side of the equation that models the problem, i.e., $\frac{110 \times 1.5 + 85 \times 3}{1.5 + 3}$, is a superposition of the one term $\frac{110 \times 1.5 + 85 \times 3}{1.5 + 3}$.

2. a. Single-term subexpressions: t_1, $-t_2$, t_3, $-t_4$

   Two-term subexpressions: $t_1 - t_2$, $t_1 + t_3$, $t_1 - t_4$, $-t_2 + t_3$, $-t_2 - t_4$, $t_3 - t_4$

   Three-term subexpressions: $t_1 - t_2 + t_3$, $t_1 - t_2 - t_4$, $t_1 + t_3 - t_4$, $-t_2 + t_3 - t_4$

   Four-term subexpression: $t_1 - t_2 + t_3 - t_4$

   b. Single-term subexpressions: $-t_1$, t_2, $-t_3$, t_4

   Two-term subexpressions: $-t_1 + t_2$, $-t_1 - t_3$, $-t_1 + t_4$, $t_2 - t_3$, $t_2 + t_4$, $-t_3 + t_4$

   Three-term subexpressions: $-t_1 + t_2 - t_3$, $-t_1 + t_2 + t_4$, $-t_1 - t_3 + t_4$, $t_2 - t_3 + t_4$

   Four-term subexpression: $-t_1 + t_2 - t_3 + t_4$

3. a. Inverse superposition
   b. Direct superposition
   c. Inverse superposition
   d. t decreases by the same amount.
   e. t increases by the same amount.
   f. t increases by the same amount.
   g. t decreases by the same amount.
   h. t decreases by the same amount.
   i. t increases by the same amount.

4. a. If an increase/decrease in q_1 by a certain amount forces an increase/decrease in q_2 by the same amount, then q_2 is directly superpositional to q_1. Since direct superposition is symmetric, q_1 is directly superpositional to q_2. This means that an increase/decrease in q_2 by a certain amount forces an increase/decrease in q_1 by the same amount.

   b. If an increase/decrease in q_1 by a certain amount forces a decrease/increase in q_2 by the same amount, then q_2 is inversely superpositional to q_1. Since inverse superposition is symmetric, q_1 is inversely superpositional to q_2. This means that an increase/decrease in q_2 by a certain amount forces a decrease/increase in q_1 by the same amount.

Exercise Set 3.3

1. a. The equation that models the problem contains three terms. The terms along with the quantities whose values they represent are listed below.

a	amount of money Marie spent ($)
1.99×3	amount of money spent on the pens ($)
12.80	amount of money spent on the notebook ($)

   The equation that models the problem states that *the amount of money Marie spent is equal to the sum of amount of money spent on the pens and amount of money spent on the notebook* or, less formally, *the amount of money that Marie spent is equal to the amount of money spent on the pens plus the amount of money spent on the notebook.*

   b. The equation that models the problem contains two terms. The terms along with the quantities whose values they represent are listed below.

m	mass of a molecule of O_3 (amu)
16.00×3	mass of a molecule of O_3 (amu)

The equation that models the problem states that *the mass of a molecule of O_3 is equal to the mass of a molecule of O_3.*[3]

c. The equation that models the problem contains three terms. The terms along with the quantities whose values they represent are listed below.

m mass of 1 mol of H_2O (g)
1.008×2 mass of 2 mol of H (g)
16.00 mass of 1 mol of O (g)

The equation that models the problem states that *the mass of 1 mol of H_2O is equal to the sum of the mass of 2 mol of H and the mass of 1 mol of O* or, less formally, *the mass of 1 mol of H_2O is equal to the mass of 2 mol of H plus the mass of 1 mol of O.*

d. The equation that models the problem contains three terms. The terms along with the quantities whose values they represent are listed below.

a amount of money Stephan has left ($)
32.80 amount of money Stephan started with ($)
$(1.30 \times 2 + 3.50 \times 4)$ amount of money Stephan spent ($)

The equation that models the problem states that *the amount of money that Stephan has left is equal to the difference between the amount of money that he started with and the amount of money that he spent.* or, less formally, *the amount of money that Stephan has left is equal to the amount of money that he started with minus the amount of money that he spent.*

e. The equation that models the problem contains four terms. The terms along with the quantities whose values they represent are listed below.

E energy content of the cup of yogurt (Cal)
4×21.6 energy contribution from carbohydrates (Cal)
9×12 energy contribution from fat (Cal)
4×10.2 energy contribution from protein (Cal)

The equation that models the problem states that *the energy content of the cup of yogurt is equal to the sum of energy contributions from carbohydrates, fat and protein* or, less formally, *the energy content of the cup of yogurt is equal to the energy contribution from carbohydrates plus the energy contribution from fat plus the energy contribution from proteins.*

f. The equation that models the problem contains three terms. The terms along with the quantities whose values they represent are listed below.

---

[3]The reading of the equation in this model along the lines of superposition as given above seems redundant. Such redundancies are seen in equations that have a single term on their left side and a single term on their right side. The difference between the two occurrences of the phrase *mass of a molecule of* O_3 is in the detail that emerges with further analysis of terms. In the example above, the occurrence of the phrase *mass of a molecule of* O_3 on the left side of the equation is represented by the symbol that represents the value of the mass of a molecule of O_3 while the occurrence of the phrase *mass of a molecule of* O_3 on the right side of the equation describes the manner in which this value can be computed if one knows the value of the molar mass of O and the value of the number of O atoms in a molecule of O_3.

V volume of salt solution left (L)

32.4   volume of salt solution at the start (L)

$(1.25 \times 2.5 + 3.25 \times 1.5)$ volume of salt solution used (L)

The equation that models the problem states that *the volume of salt solution left is equal to the difference between the volume of salt solution at the start and the volume of salt solution used* or, less formally, *the volume of salt solution left is equal to the volume of the salt solution at the start minus the volume of salt solution used.*

g. The equation that models the problem contains two terms. The terms along with the quantities whose values they represent are listed below.

m mass of 4.5 mol of C_2H_4 (g)

$4.5\,(12.01 \times 2 + 1.008 \times 4)$ mass of 4.5 mol of C_2H_4 (g)

The equation that models the problem states that *the mass of 4.5 mol of C_2H_4 is equal to the mass of 4.5 mol of C_2H_4.*

h. The equation that models the problem contains three terms. The terms along with the quantities whose values they represent are listed below.

d total distance I covered today (km)

120×1.5 distance I covered this morning (km)

90×2.5 distance I covered this afternoon (km)

The equation that models the problem states that *total distance I covered today is equal to the sum of the distances that I covered this morning and this afternoon* or, less formally, *total distance I covered today is equal to the distance I covered this morning plus the distance I covered this afternoon.*

i. The equation that models the problem contains three terms. The terms along with the quantities whose values they represent are listed below.

l length of board after the cuts (m)

1.82   initial length of the board (m)

0.35×3 total length of pieces cut from the board (m)

The equation that models the problem states that *the length of the board after the cuts is equal to the difference between the initial length of the board and the total length of the pieces that were cut from it* or, less formally, *the length of the board after the cuts is equal to the initial length of the board minus the total length of the pieces that were cut from it.*

j. The equation that models the problem contains two terms. The terms along with the quantities whose values they represent are listed below.

$\overline{m}$ average mass of a box (g/1)

$\dfrac{20.5 \times 2 + 8.5 \times 3}{2 + 3}$ average mass of a box (g/1)

The equation that models the problem states that *the average mass of a box is equal to the average mass of a box.*

k. The equation that models the problem contains four terms. The terms along with the quantities whose values they represent are listed below.

$\overline{m}$ average mass of an O atom in the sample (amu/1)
16×0.973   contribution to average mass by ^{16}O isotopes (amu/1)
17×0.026   contribution to average mass by ^{17}O isotopes (amu/1)
18×0.001   contribution to average mass by ^{18}O isotopes (amu/1)

The equation that models the problem states that *the average mass of an O atom in the sample is equal to the sum of the contributions from the ^{16}O, ^{17}O and ^{18}O isotopes* or, less formally, *the average mass of an O atom in the sample is equal to the contribution from the ^{16}O isotopes, ^{17}O isotopes and ^{18}O isotopes.*

1. The equation that models the problem contains two terms. The terms along with the quantities whose values they represent are listed below.

$\overline{v}$ my average speed today (km/h)
$\dfrac{110 \times 1.5 + 85 \times 3}{1.5 + 3}$ my average speed toady (km/h)

The equation that models the problem states that *my average speed today is equal to my average speed today.*

2. a. The energy content per cup of coffee.
   b. The number of cups of coffee Maria had.
   c. The energy content of the slice of buttered toast.
   d. Energy loss due to resting.
   e. Energy loss due to running.
   f. The energy content per egg.
   g. The number of eggs Maria had.
   h. Maria's intake of energy from the cups of coffee.
   i. Maria's intake of energy from the eggs.
   j. Maria's intake of energy from the cups of coffee and the slice of buttered toast.
   k. Maria's net intake of energy before she went running.
   l. Maria's net intake of energy after she went running.
   m. Maria's net loss of energy due to resting and running.
   n. Maria's net intake of energy from the cups of coffee, the slice of buttered toast and the eggs.

Exercise Set 3.4

1. a. The expression on the right side of the equation that models the problem is a superposition of two terms. The first term, i.e., 1.99×3, represents the cost of the pens and is a source of the expression $1.99 \times 3 + 12.80$ which represents the amount of money Marie spent. The second term, i.e., 12.80, represents the cost of the notebook and is also a source of the expression $1.99 \times 3 + 12.80$ which represents the amount of money Marie spent.

   b. The expression on the right side of the equation that models the problem is a superposition of the single term 16.00×3 which represents the mass of a molecule of O_3 and is a source of the expression 16.00×3 which represents the mass of a molecule of O_3.

c. The expression on the right side of the equation that models the problem is a superposition of two terms. The first term, i.e., 1.008×2, represents the mass of 2 mol of H and is a source of the expression $1.008 \times 2 + 16.00$ which represents the mass of 1 mol of H_2O. The second term, i.e., 16.00, represents the mass of 1 mol of O and is also a source of the expression $1.008 \times 2 + 16.00$ which represents the mass of 1 mol of H_2O.

d. The expression on the right side of the equation that models the problem is a superposition of two terms. The first term, i.e., 32.80, represents the initial amount of money that Stephan had and is a source of money that Stephan has, and the second term, i.e., $(1.30 \times 2 + 3.50 \times 4)$, represents total money spent and is a sink of money that Stephan has.

The second term is itself a superposition of two terms. The first term, i.e., 1.30×2, represents the amount of money spent on the pens and is a source of the expression $1.30 \times 2 + 3.50 \times 4$ which represents the total amount of money spent. The second term, i.e., 3.50×4, represents the amount of money spent on the notebooks and is also a source of the expression $1.30 \times 2 + 3.50 \times 4$ which represents the total amount of money spent.

e. The expression on the right side of the equation that models the problem is a superposition of three terms. The first term, i.e., 4×21.6, represents the energy contribution from the carbohydrate content of the cup of yogurt and is a source of the expression $4 \times 21.6 + 9 \times 12 + 4 \times 10.2$ which represents the energy content of the cup of yogurt. The second term, i.e., 9×12, represents the energy contribution from the fat content of the cup of yogurt and is also a source of the expression $4 \times 21.6 + 9 \times 12 + 4 \times 10.2$ which represents the energy content of the cup of yogurt. The third term, i.e., 4×10.2, represents the energy contribution from the protein content of the cup of yogurt and is a source of the expression $4 \times 21.6 + 9 \times 12 + 4 \times 10.2$ which represents the energy content of the cup of yogurt.

f. The expression on the right side of the equation that models the problem is a superposition of two terms. The first term, i.e., 32.4, represents the volume of the salt solution at the start and is a source of the expression $32.4 - (1.25 \times 2.5 + 3.25 \times 1.5)$ which represents the volume of salt solution left. The second term, i.e., $(1.25 \times 2.5 + 3.25 \times 1.5)$, represents the volume of salt solution used and is a sink of the expression $32.4 - (1.25 \times 2.5 + 3.25 \times 1.5)$ which represents the volume of salt solution left.

The second term is itself a superposition of two terms. The first, i.e., 1.25×2.5 represents the volume of the salt solution used this morning and is a source of the expression $1.25 \times 2.5 + 3.25 \times 1.5$ which represents the volume of the salt solution used. The second term, i.e., 3.25×1.5, represents the volume of the salt solution used this afternoon and is a source the expression $1.25 \times 2.5 + 3.25 \times 1.5$ which represents the volume of salt solution used.

g. The expression on the right side of the equation that models the problem is a superposition of the single term $4.5\,(12.01 \times 2 + 1.008 \times 4)$ and represents the mass of 4.5 mol of C_2H_4. This term is a source of the expression $4.5\,(12.01 \times 2 + 1.008 \times 4)$ which represents the mass of 4.5 mol of C_2H_4.

The second factor within this term, i.e., $(12.01 \times 2 + 1.008 \times 4)$, is a superposition of two terms: The first term, i.e., 12.01×2, represents the mass of 2 mol of C atoms and is a source of the expression $12.01 \times 2 + 1.008 \times 4$ which represents the mass of 1 mol of C_2H_4. The second term, i.e., 1.008×4, represents the mass of 4 mol of H atoms and is also a source of the expression $12.01 \times 2 + 1.008 \times 4$ which represents the mass of 1 mol of C_2H_4.

h. The expression on the right side of the equation that models the problem is a superposition of two terms. The first term, i.e., 120×1.5, represents distance covered this morning and is a source of the expression $120 \times 1.5 + 90 \times 2.5$ which represents the total distance covered. The second term, i.e., 90×2.5, represents distance covered this afternoon and is also a source of the expression $120 \times 1.5 + 90 \times 2.5$ which represents the total distance covered.

i. The expression on the right side of the equation that models the problem is a superposition of two terms. The first term, i.e., 1.82, represents the initial length of the board and is a source of the expression $1.82 - 0.35 \times 3$ which represents the remaining length. The second term, i.e., 0.35×3, represents the total length of the pieces that were cut from the board and is a sink of the expression $1.82 - 0.35 \times 3$ which represents the remining length.

j. The expression on the right side of the equation that models the problem is a superposition of the single term $\frac{20.5 \times 2 + 8.5 \times 3}{2+3}$ which represents the average mass of a box and is a source of the expression $\frac{20.5 \times 2 + 8.5 \times 3}{2+3}$ which represents the average mass of a box.

The numerator of the expression on the right side of the equation that models the problem is a superposition of two terms. The first term, i.e., 20.5×2, represents the mass of the 2 boxes and is a source of the expression $20.5 \times 2 + 8.5 \times 3$ which represents the mass of all the boxes. The second term, i.e., 8.5×3, represents the mass of the 3 boxes and is also a source of the expression $20.5 \times 2 + 8.5 \times 3$ which represents the mass of all the boxes.

The denominator of the expression on the right side of the equation that models the problem is a superposition of two terms as well. The first term, i.e., 2, represents the number of boxes of mass 20.5 g and is a source of the expression $2 + 3$ which represents the total number of boxes. The second term, i.e., 3, represents the number of boxes of mass 8.5 g and is also a source of the expression $2 + 3$ which represents the total number of boxes.

k. The expression on the right side of the equation that models the problem is a superposition of three terms. The first term, i.e., 16×0.973, represents the contribution to the average mass by the ^{16}O isotopes and is a source of the expression $16 \times 0.973 + 17 \times 0.026 + 18 \times 0.001$ which represents the average mass of an O atom in the sample. The second term, i.e., 17×0.026, represents the contribution to the average mass by the ^{17}O isotopes and is also a source of the expression $16 \times 0.973 + 17 \times 0.026 + 18 \times 0.001$ which represents the average mass of an O atom in the sample. The third term, i.e., 18×0.001, represents the contribution to the average mass by the ^{18}O

isotopes and is a source of the expression $16 \times 0.973 + 17 \times 0.026 + 18 \times 0.001$ which represents the average mass of an O atom in the sample.

1. The expression on the right side of the equation that models the problem is a superposition of the single term $\frac{110 \times 1.5 + 85 \times 3}{1.5 + 3}$ which represents my average speed today and is a source of the expression $\frac{110 \times 1.5 + 85 \times 3}{1.5 + 3}$ which represents my average speed today.

The numerator of the expression on the right side of the equation that models the problem is a superposition of two terms. The first term, i.e., 110×1.5, represents distance covered this morning and is a source of the expression $110 \times 1.5 + 85 \times 3$ which represents the total distance covered today. The second term, i.e., 85×3, represents distance covered this afternoon and is also a source of the expression $110 \times 1.5 + 85 \times 3$ which represents the total distance covered today.

The denominator of the expression on the right side of the equation that models the problem is a superposition of two terms as well. The first term, i.e., 1.5, represents the duration of driving this morning and is a source of the expression $1.5 + 3$ which represents the total travel time today. The second term, i.e., 3, represents the duration of driving this afternoon and is also a source of the expression $1.5 + 3$ which represents the total travel time today.

Exercise Set 3.5

1. a. We list the terms and work out their units.

> Term: a
> Unit: $

> Term: 1.99×3
> Unit: $/1×1 = $ or, less formally, $/pen×pens = $

> Term: 12.80
> Unit: $

Since all terms have the same unit, the equation is consistent.

b. We list the terms and work out their units.

> Term: m
> Unit: amu

> Term: 16.00×3
> Unit: amu/1×1 = amu or, less formally, amu/atom×atoms = amu

Since all terms have the same unit, the equation is consistent.

c. We list the terms and work out their units.

> Term: m
> Unit: g

Term: 1.008×2
Unit: $\text{g/mol} \times \text{mol} = \text{g}$

Term: 16.00
Unit: g

Since all terms have the same unit, the equation is consistent.

d. We list the terms and work out their units.

Term: a
Unit: $

Term: 32.80
Unit: $

Term: $1.30 \times 2 + 3.50 \times 4$
Unit: $\text{\$}/1 \times 1 + \text{\$}/1 \times 1 = \text{\$} + \text{\$}$
$$= \text{\$}$$

or, less formally,

$\text{\$}/\text{pen} \times \text{pens} + \text{\$}/\text{notebook} \times \text{notebooks} = \text{\$} + \text{\$}$
$$= \text{\$}$$

Since all terms have the same unit, the equation is consistent.[4]

e. We list the terms and work out their units.

Term: E
Unit: Cal

Term: 4×21.6
Unit: $\text{Cal/g} \times \text{g} = \text{Cal}$

Term: 9×12
Unit: $\text{Cal/g} \times \text{g} = \text{Cal}$

Term: 4×10.2
Unit: $\text{Cal/g} \times \text{g} = \text{Cal}$

Since all terms have the same unit, the equation is consistent.

f. We list the terms and work out their units.

Term: V
Unit: L

Term: 32.4
Unit: L

---

[4] Here is an example of a difference between the rules of the algebra of quantities and the algebra of units. Note that, while the expression $m + m$ is equivalent to $2m$ (with m representing the value of some mass), the expression m+m is equivalent to m (with m representing the size of the unit *metre*.

Term: $1.25 \times 2.5 + 3.25 \times 1.5$
Unit: $L/h \times h + L/h \times h = L + L$
$\qquad\qquad\qquad\qquad\quad = L$

Since all terms have the same unit, the equation is consistent.

g. We list the terms and work out their units.

Term: m
Unit: g

Term: $4.5\,(12.01 \times 2 + 1.008 \times 4)$
Unit: $mol_{C_2H_4} \times (g/mol_C \times mol_C/mol_{C_2H_4} + g/mol_H \times mol_H/mol_{C_2H_4})$
$\qquad = mol_{C_2H_4} \times (g/mol_{C_2H_4} + g/mol_{C_2H_4})$
$\qquad = mol_{C_2H_4} \times g/mol_{C_2H_4}$
$\qquad = g$

Since all terms have the same unit, the equation is consistent.

h. We list the terms and work out their units.

Term: d
Unit: km

Term: 120×1.5
Unit: $km/h \times h = km$

Term: 90×2.5
Unit: $km/h \times h = km$

Since all terms have the same unit, the equation is consistent.

i. We list the terms and work out their units.

Term: l
Unit: m

Term: 1.82
Unit: m

Term: 0.35×3
Unit: $m/1 \times 1 = m$ or, less formally, $m/piece \times pieces = m$

Since all terms have the same unit, the equation is consistent.

j. We list the terms and work out their units.

Term: $\overline{m}$
Unit: $g/1$ or, less formally, g/box

Term: $\dfrac{20.5 \times 2 + 8.5 \times 3}{2 + 3}$
Unit: $\dfrac{g/1 \times 1 + g/1 \times 1}{1 + 1} = \dfrac{g + g}{1}$
$\qquad\qquad\qquad\qquad = \dfrac{g}{1}$
$\qquad\qquad\qquad\qquad = g/1$

or, less formally,

$$\frac{\text{g/box} \times \text{boxes} + \text{g/box} \times \text{boxes}}{\text{box} + \text{box}} = \frac{\text{g} + \text{g}}{\text{box}}$$
$$= \frac{\text{g}}{\text{box}}$$
$$= \text{g/box}$$

Since all terms have the same unit, the equation is consistent.

k. We list the terms and work out their units.

Term: $\overline{m}$
Unit: amu/1 or, less formally, amu/atom

Term: 16×0.973
Unit: $\text{amu}/1 \times 1/1 = \text{amu}/1$

Term: 17×0.026
Unit: $\text{amu}/1 \times 1/1 = \text{amu}/1$

Term: 18×0.001
Unit: $\text{amu}/1 \times 1/1 = \text{amu}/1$

Since all terms have the same unit, the equation is consistent.

l. We list the terms and work out their units.

Term: $\overline{v}$
Unit: km/h

Term: $\dfrac{110 \times 1.5 + 85 \times 3}{1.5 + 3}$

Unit: $\dfrac{\text{km/h} \times \text{h} + \text{km/h} \times \text{h}}{\text{h} + \text{h}} = \dfrac{\text{km} + \text{km}}{\text{h}}$
$$= \dfrac{\text{km}}{\text{h}}$$
$$= \text{km/h}$$

Since all terms have the same unit, the equation is consistent.

Exercise Set 3.6

1.

Δp change in the value of pressure (Pa)
$\Delta p = 323\,000 - 320\,000$

$\Delta p = 3000$ Pa

The pressure changed by 3000 Pa or the pressure increased by 3000 Pa.

2.

Δd change in the value of Simon's distance from Toronto (km)
$\Delta d = 345 - 120$

$\Delta d = 225$ km

The change in the value of Simon's distance from Toronto is 225 km or Simon's distance from Toronto increased by 225 km.

3.

Δt change in the value of temperature (°C)
$\Delta t = -6.5 - 2.5$

$\Delta t = -9$ °C

The temperature changed by -9 °C or the temperature dropped by 9 °C.

4.

ΔV change in the value of volume (cm^3)
$\Delta V = 9110 - 8420$

$\Delta V = 690$ cm^3

The volume changed by 690 cm^3 or the volume increased by 690 cm^3.

5.

Δn change in the value of amount of helium (mol)
$\Delta n = 3.5 - 4.75$

$\Delta n = -1.25$ mol

The amount of helium changed by -1.25 mol or the amount of helium decreased by 1.25 mol.

6.

Δt change in the value of temperature (°C)
$\Delta t = -5.9 - (-15.8)$

$\Delta t = -5.9 + 15.8$
$\Delta t = 9.9$ °C

The temperature changed by 9.9 °C or the temperature rose by 9.9 °C.

7.

ΔE change in the value of gravitational potential energy (J)
ΔE = 1350 − (−2400)

ΔE = 1350 + 2400
ΔE = 3750 J

The gravitational potential energy of the object changed by 3750 J or the gravitational potential energy of the object rose by 3750 J.

8.

ΔI change in the value of current through the wire (A)
ΔI = 0.055 − 0.083

ΔI = −0.028 A

The current through the wire changed by − 0.028 A or the current throught the wire decreased by 0.028 A.

9. a. A positive value of Δq implies that q increased.
   b. A negative value of Δq implies that q decreased.

10. No. As an example, a change of 5 °C in the value of temperature may relate to an initial value of 20 °C and a final value of 25 °C, an initial value of −2 °C and a final value of 3 °C, or any other pair of initial and final values where the final value is 5 units larger than the initial value.

11. a. No. As an example, a change of 5 °C in the value of temperature may relate to an initial value of 20 °C and a final value of 25 °C, an initial value of −2 ° and a final value of 3 °C, or an initial value of −8 °C and a final value of −3 °C.
    b. No. As an example, a change of −5 °C in the value of temperature may relate to an initial value of 25 °C and a final value of 20 °C, an initial value of 3 ° and a final value of −2 °C, or an initial value of −3 °C and a final value of −8 °C.

Exercise Set 3.7

1. Since Francesca's net annual income is fixed, the quantity symbol a on the left side of the equation that models the problem is constant. An increase in Francesca's expenses by $1400/year increases the value of the first term by $1400/year. The only way for a to remain constant is for the value of the second term to decrease by $1400/year. This implies that Francesca's annual savings will drop by $1400/year.

2. Since the distance from Toronto to Montreal is fixed, the quantity symbol d on the left side of the equation that models the problem is constant. Reducing the remaining distance implies a reduction in the value of the second term in the expression on the right side of the model. This would force an increase in the value of the first term which represents the distance that I have covered so far. Distance covered so far could have been increased if I had driven faster or longer or both.

3. Since the momentum of the system is conserved, the quantity symbol p on the left side of the equation that models the problem is constant. A decrease in the value of the momentum of the first ball by 0.12 kg·m/s implies a reduction in the value of the first term by 0.12 kg·m/s. To keep p constant, the value of the second term will have to increase by 0.12 kg·m/s. This implies an increase in the value of the momentum of the second ball by 0.12 kg·m/s.

Exercise Set 3.8

1. a. The percentages can be added and subtracted as they both refer to parts of the same quantity, namely, *the school budget.*

b. The percentages can be added and subtracted as they both refer to parts of the same quantity, namely, *the number of daycare centres in Toronto.*

c. The percentages cannot be added and subtracted as they refer to parts of different quantities, namely, *the number of tomatoes* and *the number of bananas.*

d. The percentages can be added and subtracted as they both refer to parts of the same quantity, namely, *the workforce.*

e. The percentages cannot be added and subtracted as they refer to parts of different quantities, namely, *the price* and *the sale price.*

f. The percentages can be added and subtracted as they both refer to parts of the same quantity, namely, *the volume of the acid solution.*

g. The percentages cannot be added and subtracted as they refer to parts of different quantities, namely, *the mass of the salt* and *the mass of the sugar.*

h. The percentages cannot be added and subtracted as they refer to parts of different quantities, namely, *the number of students in Section A* and *the number of students in Section B.*

i. The percentages can be added and subtracted as they both refer to parts of the same quantity, namely, *the amount of my income.*

j. The percentages cannot be added and subtracted as they refer to parts of different quantities, namely, *the number of eligible voters* and *the number of those who voted.*

k. The percentages can be added and subtracted as they refer to parts of the same quantity, namely, *the number of turtles in the sample.*

2. a.

r percentage of students who got at least a B− (1)
$r = 6\% + 17\%$

$r = 23\%$

23% of the students got at least a B−.

b.

r percentage of students who got a grade in the B range (1)
$r = 38\% - 12\%$

$r = 26\%$

26% of the students got a grade in the B range.

c.

r percentage of my income left after expenses (1)
$r = 100\% - (40\% + 25\% + 15\%)$

$r = 100\% - 80\%$
$r = 20\%$

20% of my income is left after expenses.

d.

r percentage of non-full-time employees at the ABC plant (1)
$r = 100\% - 73\%$

$r = 27\%$

27% of the employees at the ABC plant are non-full-time.

e.

r percentage of the the project that has been completed (1)
$r = 40\% + 25\% + 20\%$

$r = 85\%$

85% of the project has been completed.

f.

r percentage of the project that has not been completed (1)
$r = 100\% - (20\% + 10\%)$

$r = 100\% - 30\%$
$r = 70\%$

70% of the project has not been completed.

3. a. The argument is invalid. The percentages cannot be added or subtracted because they refer to parts of different quantities, namely, *the number of employees in Section A* and *the number of employees in Section B*.

   b. The argument is valid. The percentages can be added or subtracted as they refer to parts of the same quantity, namely *the amount of helium in the balloon.*

c. The argument is invalid. The percentages cannot be added or subtracted because they refer to parts of different quantities, namely, *the number of crops in the first field* and *the number of crops in the second field*.[5]

d. The argument is invalid. The percentages cannot be added or subtracted because they refer to parts of different quantities, namely, *the number of frogs in the sample* and *the number of green-skinned frogs in the sample*.

e. The argument is valid. The percentages can be added or subtracted as they refer to parts of the same quantity, namely *the number of customers*.

f. The argument is invalid. The percentages cannot be added or subtracted because they refer to parts of different quantities, namely, *the number of eligible voters*, *the number of eligible voters who voted*, and *the number of eligible voters who voted and said that they support a strike if changes are implemented*.

Exercise Set 4

1. It means that q_1 and q_2 both increase or both decrease in value.

2. It means that q_1 and q_2 both increase or both decrease in value and that they do so by equal amounts.

3. Yes, it is possible for q_2 to be directly related to q_1 but not be directly superpositional to q_1. This happens if q_1 and q_2 both increase or both decrease (so that the relationship is direct) but not by equal amounts (so that the relationship is not superpositional).

4. a. V is directly superpositional to the terms 32.7 and 12.5×3.

   b. E is directly superpositional to the terms 88×2, 138 and 120×2.

   c. Δt is directly superpositional to the term 18.9.

   d. a is directly superpositional to the term 45 000.

   e. a is directly superpositional to the term 102.56.

   f. m is directly superpositional to the term 2.8×135.

   g. V is directly superpositional to the term 175.

5. a. The terms 32.7 and 12.5×3 are sources of V. This means that *the volume of solution that Mina started with* and *the volume of the solution delivered* are sources of *the volume of the solution Mina has now*.

   b. The terms 88×2, 138 and 120×2 are sources of E. This means that *intake of energy from the slices of toast*, *intake of energy from the coffee* and *intake of energy from the eggs* are sources of *Marco's net intake of energy this morning*.

   c. The term 18.9 is a source of ΔT. This means that *the value of the temperature this morning* is a source of *change in the value of the temperature*.

   d. The term 45000 is a source of a. This means that *Enrique's net annual income* is a source of *the amount of money Enrique saves every year*.

---

[5]Note also that at most 100% of the crop can get affected by adverse weather conditions.

e. The term 102.56 is a source of a. This means that *the amount of money Phillip started with* is a source of *the amount of money Phillip has left*.

f. The term 2.8×135 is a source of m. This means that *the mass of recycling materials delivered* is a source of *the mass of recycling material waiting to be processed*.

g. The term 175 is a source of V. This means that *the volume of the swimming pool* is a source of *the volume of water remaining*.

6. $-q_1, q_2, -q_3, q_4, -q_1 + q_2, -q_1 - q_3, -q_1 + q_4, q_2 - q_3, q_2 + q_4, -q_3 + q_4,$
   $-q_1 + q_2 - q_3, -q_1 + q_2 + q_4, -q_1 - q_3 + q_4, q_2 - q_3 + q_4, -q_1 + q_2 - q_3 + q_4$

Exercise Set 5

1. It means that an increase in q_1 forces a decrease in q_2 and a decrease in q_1 forces an increase in q_2.

2. It means that an increase in q_1 by a certain amount forces a decrease in q_2 by the same amount and a decrease in q_1 by a certain amount forces an increase in q_2 by the same amount.

3. Yes, it is possible for q_2 to be inversely related to q_1 but not be inversely superpositional to q_1. This happens if an increase/decrease in q_1 forces a decrease/increase in q_2 (so that the relationship is inverse) but not by equal amounts (so that the relationship is not superpositional).

4. a. There are no terms to which V is inversely superpositional.

   b. E is inversely superpositional to the term 187×1.5.

   c. Δt is inversely superpositional to the term 15.7.

   d. a is inversely superpositional to the term 38 500.

   e. a is inversely superpositional to the term $(2.30 \times 2 + 5.99 \times 3)$.

   f. m is inversely superpositional to the term 280.

   g. V is inversely superpositional to the term $(35 \times 0.5 + 40 \times 0.25)$.

5. a. There are no sinks of V.

   b. The term 187×1.5 is a sink of E. This means that *loss of energy due to exercising* is a sink of *Marco's net intake of energy this morning*.

   c. The term 15.7 is a sink of Δt. This means that *the value of the temperature this morning* is a sink of *change in the value of the temperature*.

   d. The term 38 500 is a sink of a. This means that *Enrique's annual expenses* is a sink of *the amount of money Enrique saves every year*.

   e. The term $(2.30 \times 2 + 5.99 \times 3)$ is a sink of a. This means that *the amount of money Phillip spent* is a sink of *the amount of money Phillip has left*.

   f. The term 280 is a sink of m. This means that *the mass of recycling material that has been processed* is a sink of *the mass of recycling material waiting to be processed*.

   g. The term $(35 \times 0.5 + 40 \times 0.25)$ is a sink of V. This means that *the volume of water that has been removed* is a sink of *the volume of water remaining*.

6. $q_1, -q_2, q_3, -q_4, q_1 - q_2, q_1 + q_3, q_1 - q_4, -q_2 + q_3, -q_2 - q_4, q_3 - q_4, q_1 - q_2 + q_3,$
   $q_1 - q_2 - q_4, q_1 + q_3 - q_4, -q_2 + q_3 - q_4, q_1 - q_2 + q_3 - q_4$

Exercise Set 6.1

1. a. The expression on the right side of the equation that models the problem, i.e., $0.87 \times 2 + 7.20 \times 3$, is a superposition of the two terms 0.87×2 and 7.20×3. The first term is a proportion involving the two factors 0.87 and 2, and the second term is a proportion involving the two factors 7.20 and 3.

   b. The expression on the right side of the equation that models the problem, i.e., 1.008×2, is a superposition of the one term 1.008×2. This term is a proportion involving the factors 1.008 and 2.

   c. The expression on the right side of the equation that models the problem, i.e., $12.01 + 16.00 \times 2$, is a superposition of the two terms 12.01 and 16.00×2. The first term is a proportion involving the single factor 12.01. The second term is a proportion involving the two factors 16.00 and 2.

   d. The expression on the right side of the equation that models the problem, i.e., $41.83 - (1.10 \times 3 + 6.25 \times 4)$, is a superposition of the two terms 41.83 and $(1.10 \times 3 + 6.25 \times 4)$. The first term is a proportion involving the single factor 41.83. The second term analyzes into the single factor $(1.10 \times 3 + 6.25 \times 4)$ which further analyzes into the two terms 1.10×3, which is a proportion involving the two factors 1.10 and 3, and 6.25×4 which is a proportion involving the two factors 6.25 and 4.

   e. The expression on the right side of the equation that models the problem, i.e., $4 \times 11.4 + 9 \times 4.5 + 4 \times 7.8$, is a superposition of the three terms 4×11.4, 9×4.5 and 4×7.8. The first term, i.e., 4×11.4, is a proportion involving the two factors 4 and 11.4. The second term, i.e., 9×4.5, is a proportion involving the two factors 9 and 4.5. The third term, i.e., 4×7.8, is a proportion involving the two factors 4 and 7.8.

   f. The expression on the right side of the equation that models the problem, i.e., $41.8 - (2.1 \times 1.5 + 4.7 \times 2)$, is a superposition of the two terms 41.8 and $(2.1 \times 1.5 + 4.7 \times 2)$. The first term is a proportion involving the single factor 41.8. The second term is a proportion involving the single factor $(2.1 \times 1.5 + 4.7 \times 2)$ which is a superposition of the two terms 2.1×1.5 and 4.7×2. The first term, i.e., 2.1×1.5, is a proportion involving the two factors 2.1 and 1.5. The second term, i.e., 4.7×2, is a proportion involving the two factors 4.7 and 2.

   g. The expression on the right side of the equation that models the problem, i.e., $2.3\,(12.01 \times 3 + 1.008 \times 8)$, is a superposition of the one term $2.3\,(12.01 \times 3 + 1.008 \times 8)$. The single term $2.3\,(12.01 \times 3 + 1.008 \times 8)$ is a proportion involving the two factors 2.3 and $(12.01 \times 3 + 1.008 \times 8)$ the second of which is a superposition of the two terms 12.01×3 and 1.008×8. The first term, i.e., 12.01×3, is a proportion involving the two factors 12.01 and 3. The second term, i.e., 1.008×8, is a proportion involving the two factors 1.008 and 8.

   h. The expression on the right side of the equation that models the problem, i.e., $110 \times 1.5 + 80 \times 2.5$, is a superposition of the two terms 110×1.5 and 80×2.5. The first term, i.e., 110×1.5, is a proportion involving the two factors 110 and 1.5. The second term, i.e., 80×2.5, is a proportion involving the two factors 80 and 2.5.

i. The expression on the right side of the equation that models the problem, i.e., $\frac{11.9 \times 3 + 7.1 \times 2}{3 + 2}$, is a superposition of the one term $\frac{11.9 \times 3 + 7.1 \times 2}{3 + 2}$. The single term $\frac{11.9 \times 3 + 7.1 \times 2}{3 + 2}$ is a proportion involving the single factor $\frac{11.9 \times 3 + 7.1 \times 2}{3 + 2}$. It may also be described as a proportion involving the factor $(11.9 \times 3 + 7.1 \times 2)$ in the dividend and the factor $(3 + 2)$ in the divisor. The factor in the dividend, i.e., $(11.9 \times 3 + 7.1 \times 2)$, is a superposition of the two terms 11.9×3 which is a proportion involving the two factors and 11.9 and 3, and 7.1×2 which is a proportion involving the two factors 7.1 and 2. The factor in the divisor, i.e., $(3 + 2)$ is a superposition of the two terms 2 and 3.

j. The expression on the right side of the equation that models the problem, i.e., $16 \times 0.973 + 17 \times 0.026 + 18 \times 0.001$, is a superposition of the three terms 16×0.973 which is a proportion involving the two factors 16 and 0.973, 17×0.026 which is a proportion involving the two factors 17 and 0.026, and 18×0.001 which is proportion involving the two factors 18 and 0.001.

k. The expression on the right side of the equation that models the problem, i.e., $\frac{110 \times 1.5 + 85 \times 3}{1.5 + 3}$, is a superposition of the one term $\frac{110 \times 1.5 + 85 \times 3}{1.5 + 3}$. The single term $\frac{110 \times 1.5 + 85 \times 3}{1.5 + 3}$ is a proportion involving the single factor $\frac{110 \times 1.5 + 85 \times 3}{1.5 + 3}$. It may also be described as a proportion involving the factor $(110 \times 1.5 + 85 \times 3)$ in the dividend and the factor $(1.5 + 3)$ in the divisor. The factor in the dividend, i.e., $(110 \times 1.5 + 85 \times 3)$, is a superposition of the two terms 110×1.5 which is a proportion involving the two factors 110 and 1.5, and 85×3 which is a proportion involving the two factors 85 and 3. The factor in the divisor, i.e., $(1.5 + 3)$ is a superposition of the two terms 1.5 and 3.

Exercise Set 6.2

1. a. The expression on the right side of the equation that models the problem, i.e., $0.87 \times 2 + 7.20 \times 3$, is a superposition of the two terms 0.87×2 which is a proportion involving the two factors 0.87 and 2, and 7.20×3 which is a proportion involving the two factors 7.20 and 3.

b. The expression on the right side of the equation that models the problem, i.e., 1.008×2, is a superposition involving the single term 1.008×2 which is a proportion involving the two factors 1.008 and 2.

c. The expression on the right side of the equation that models the problem, i.e., $12.01 + 16.00 \times 2$, is a superposition of the two terms 12.01 which is a proportion involving the single factor 12.01, and 16.00×2 which is a proportion involving the two factors 16.00 and 2.

d. The expression on the right side of the equation that models the problem, i.e., $41.83 - (1.10 \times 3 + 6.25 \times 4)$, is a superposition of the two terms 41.83 which is a proportion involving the single factor 41.83, and $(1.10 \times 3 + 6.25 \times 4)$

which is a superposition of the two terms 1.10×3 which is in turn a proportion involving the two factors 1.10 and 3, and 6.25×4 which is a proportion involving the factors 6.25 and 4.

e. The expression on the right side of the equation that models the problem, i.e., $4 \times 11.4 + 9 \times 4.5 + 4 \times 7.8$, is a superposition of the three terms 4×11.4 which is proportion involving the factors 4 and 11.4, 9×4.5 which is a proportion involving the factors 9 and 4.5, and 4×7.8 which is a proportion involving the two factors 4 and 7.8.

f. The expression on the right side of the equation that models the problem, i.e., $41.8 - (2.1 \times 1.5 + 4.7 \times 2)$, is a superposition of the two terms 41.8 which is a proportion involving the single factor 41.8, and $(2.1 \times 1.5 + 4.7 \times 2)$ which is a superposition involving the two terms 2.1×1.5 which is in turn a proportion involving the factors 2.1 and 1.5, and 4.7×2 which is a proportion involving the factors 4.7 and 2.

g. The expression on the right side of the equation that models the problem, i.e., $2.3\,(12.01 \times 3 + 1.008 \times 8)$, is a superposition of the one term $2.3\,(12.01 \times 3 + 1.008 \times 8)$ which is a proportion involving the two factors 2.3 and $(12.01 \times 3 + 1.008 \times 8)$ the latter of which is a superposition involving the two terms 12.01×3 which is a proportion involving the two factors 12.01 and 3, and 1.008×8 which is a proportion involving the two factors 1.008 and 8.

h. The expression on the right side of the equation that models the problem, i.e., $110 \times 1.5 + 80 \times 2.5$, is a superposition of the two terms 110×1.5 which is a proportion involving the two factors 110 and 1.5, and 80×2.5 which is a proportion involving the two factors 80 and 2.5.

i. The expression on the right side of the equation that models the problem, i.e., $\frac{11.9 \times 3 + 7.1 \times 2}{3+2}$, is a superposition of the one term $\frac{11.9 \times 3 + 7.1 \times 2}{3+2}$ which may be seen as a proportion involving the single factor $\frac{11.9 \times 3 + 7.1 \times 2}{3+2}$ or a proportion involving the two factors $(11.9 \times 3 + 7.1 \times 2)$ and $(3 + 2)$.

j. The expression on the right side of the equation that models the problem, i.e., $16 \times 0.973 + 17 \times 0.026 + 18 \times 0.001$, is a superposition of the three terms 16×0.973 which is a proportion involving the two factors 16 and 0.973, 17×0.026 which is a proportion involving the two factors 17 and 0.026, and 18×0.001 which is a proportion involving the two factors 18 and 0.001.

k. The expression on the right side of the equation that models the problem, i.e., $\frac{110 \times 1.5 + 85 \times 3}{1.5+3}$, is a superposition of the one term $\frac{110 \times 1.5 + 85 \times 3}{1.5+3}$ which may be seen as a proportion involving the single factor $\frac{110 \times 1.5 + 85 \times 3}{1.5+3}$ or a proportion involving the two factors $(110 \times 1.5 + 85 \times 3)$ and $(1.5 + 3)$.

2. a. Single-factor subexpressions: f_1, f_2, $\frac{1}{f_3}$, $\frac{1}{f_4}$

   Two-factor subexpressions: $f_1 f_2$, $\frac{f_1}{f_3}$, $\frac{f_1}{f_4}$, $\frac{f_2}{f_3}$, $\frac{f_2}{f_4}$, $\frac{1}{f_3 f_4}$

   Three-factor subexpressions: $\frac{f_1 f_2}{f_3}$, $\frac{f_1 f_2}{f_4}$, $\frac{f_1}{f_3 f_4}$, $\frac{f_2}{f_3 f_4}$

   Four-factor subexpression: $\frac{f_1 f_2}{f_3 f_4}$

   b. Single-factor subexpressions: $\frac{1}{f_1}$, $\frac{1}{f_2}$, f_3, f_4

   Two-factor subexpressions: $\frac{1}{f_1 f_2}$, $\frac{f_3}{f_1}$, $\frac{f_4}{f_1}$, $\frac{f_3}{f_2}$, $\frac{f_4}{f_2}$, $f_3 f_4$

Three-factor subexpressions: $\frac{f_3}{f_1 f_2}$, $\frac{f_4}{f_1 f_2}$, $\frac{f_3 f_4}{f_1}$, $\frac{f_3 f_4}{f_2}$

Four-factor subexpression: $\frac{f_3 f_4}{f_1 f_2}$

3. a. direct proportion
   b. inverse proportion
   c. inverse proportion
   d. f gets scaled by the same factor.
   e. f gets scaled by the inverse of the same factor.
   f. f gets scaled by the inverse of the same factor.
   g. f gets multiplied by the same value.
   h. f gets divided by the same value.
   i. f gets divided by the same value.
   j. f gets multiplied by the same value.
   k. f gets divided by the same value.
   l. f gets multiplied by the same value.

4. The expression

$$\frac{f_1}{f_2 f_3} + \frac{1}{f_4}$$

analyzes into the two terms $\frac{f_1}{f_2 f_3}$ and $\frac{1}{f_4}$. This is shown below:

$$\overset{t_1}{\boxed{\frac{f_1}{f_2 f_3}}} + \overset{t_2}{\boxed{\frac{1}{f_4}}}$$

In this expression

- t_1 is directly proportional to f_1. This means that scaling f_1 by a certain factor forces the scaling of t_1 by the same factor.

- t_1 is inversely proportional to each of f_2 and f_3. This means that scaling either one of f_2 or f_3 by a certain factor forces the scaling of t_1 by the inverse of the same factor.

- t_2 is inversely proportional to f_4. This means that scaling f_4 by a certain factor forces the scaling of t_2 by the inverse of the same factor.

5. a. If scaling q_1 by a certain factor forces the scaling of q_2 by the same factor, then q_2 is directly proportional to q_1. Since direct proportion is symmetric, q_1 is directly proportional to q_2. This means that scaling q_2 by a certain factor forces the scaling of q_1 by the same factor.

   b. If scaling q_1 by a certain factor forces the scaling of q_2 by the inverse of the same factor, then q_2 is inversely proportional to q_1. Since inverse proportion is symmetric, q_1 is inversely proportional to q_2. This means that scaling q_2 by a certain factor forces the scaling of q_1 by the inverse of the same factor.

Exercise Set 6.3

1. a. The total amount of money spent (the expression $0.87 \times 2 + 7.20 \times 3$) is equal to the sum of the amount of money spent on the pens (the term

0.87×2) and the amount of money spent on the notebooks (the term 7.20×3). Furthermore, we can say that the total amount of money spent is directly superpositional to the amount of money spent on the pens and the amount of money spent on the notebooks.

The amount of money spent on the pens (the term 0.87×2) is equal to the product of the cost of a pen (the factor 0.87) and the number of pens purchased (the factor 2). Furthermore, we can say that the amount of money spent on the pens is directly proportional to the cost of a pen and the number of pens purchased.

The amount of money spent on the notebooks (the term 7.20×3) is equal to the product of the cost of a notebook (the factor 7.20) and the number of notebooks purchased (the factor 3). Furthermore, we can say that the amount of money spent on the notebooks is directly proportional to the cost of a notebook and the number of notebooks purchased.

b. Since there is a single term in the expression on the right side of the model, the relationship between the expression 1.008×2 and the single term that it analyzes into, i.e., 1.008×2, is trivial: The mass of 1 mol of H_2 (the expression 1.008×2) is equal to the mass of 1 mol of H_2 (the term 1.008×2). We can also state the obvious implication that the mass of 1 mol of H_2 is directly superpositional to the mass of 1 mol of H_2.[6]

The mass of 1 mol of H_2 (the term 1.008×2) is equal to the product of the molar mass of H (the factor 1.008) and the amount of H atoms (the factor 2). Furthermore, we can say that the mass of a molecule of H_2 is directly proportional to the molar mass of H and the amount of H.

c. The mass of a molecule of CO_2 (the expression $12.01 + 16.00 \times 2$) is equal to the sum of the mass of an atom of C (the term 12.01) and the mass of 2 atoms of O (the term 16.00×2). Furthermore, we can say that the mass of a molecule of CO_2 is directly superpositional to the mass of the C atom in a molecule of CO_2 and the mass of the O atoms in a molecule of CO_2.

The mass of O atoms in a molecule of CO_2 (the term 16.00×2) is equal to the product of the atomic mass of O (the factor 16.00) and the number of O atoms in a molecule of CO_2 (the factor 2). Furthermore, we can say that the mass of O atoms in a molecule of CO_2 is directly proportional to the atomic mass of O and the number of O atoms in a molecule of CO_2.

d. The amount of money I have left (the expression $41.83 - (1.10 \times 3 + 6.25 \times 4)$) is equal to the difference between the amount of money I started with (the term 41.83) and the amount of money I spent (the term $(1.10 \times 3 + 6.25 \times 4)$). Furthermore, we can say that the amount of money I have left is directly superpositional to the amount of money I started with and inversely superpositional to the amount of money I spent.

---

[6]Such trivial relationships are seen in expressions that analyze into a single term and terms that analyze into a single factor. In what follows, we will no longer take note of such trivial relationships.

The amount of money I spent (the term $(1.10 \times 3 + 6.25 \times 4)$) is equal to the sum of the amount of money I spent on the pens (the term 1.10×3) and the amount of money I spent on the notebooks (the term 6.25×4). Furthermore, we can say that the amount of money I spent is directly superpositional to the amount of money I spent on the pens and the amount of money I spent on the notebooks.

The amount of money I spent on the pens (the term 1.10×3) is equal to the product of the cost of a pen (the factor 1.10) and the number of pens purchased (the factor 3). Furthermore, we can say that the amount of money I spent on the pens is directly proportional to the cost of a pen and the number of pens purchased.

The amount of money I spent on the notebooks (the term 6.25×4) is equal to the product of the cost of a notebook (the factor 6.25) and the number of notebooks purchased (the factor 4). Furthermore, we can say that the amount of money I spent on the notebooks is directly proportional to the cost of a notebook and the number of notebooks purchased.

e. The energy content of the cup of yogurt (the expression $4 \times 11.4 + 9 \times 4.5 + 4 \times 7.8$) is equal to the sum of the energy contribution from the carbohydrate content of the cup of yogurt (the term 4×11.4), the energy contribution from the fat content of the cup of yogurt (the term 9×4.5) and the energy contribution from the protein content of the cup of yogurt (the term 4×7.8). Furthermore, we can say that the energy content of the cup of yogurt is directly superpositional to the energy contribution from the carbohydrate content of the cup of yogurt, the energy contribution from the fat content of the cup of yogurt, and the energy contribution from the protein content of the cup of yogurt.

The energy contribution from the carbohydrate content of the cup of yogurt (the term 4×11.4) is equal to the product of the specific energy content of carbohydrates[7] and the mass of carbohydrates in the cup of yogurt. Furthermore, we can say that the energy contribution from the carbohydrate content of the cup of yogurt is directly proportional to the specific energy content of carbohydrates and the mass of carbohydrates in the cup of yogurt.

The energy contribution from the fat content of the cup of yogurt (the term 9×4.5) is equal to the product of the specific energy content of fats and the mass of fat in the cup of yogurt. Furthermore, we can say that the energy contribution from the fat content of the cup of yogurt is directly proportional to the specific energy content of fat and the mass of fat in the cup of yogurt.

---

[7] the adjective *specific* describes the value of a quantity per unit mass of the entity to which the quantity belongs. As an example, the specific energy content of carbohydrates refers to the energy that can be extracted from the unit mass of carbohydrates.

The energy contribution from the protein content of the cup of yogurt (the term 4×7.8) is equal to the product of the specific energy content of proteins and the mass of protein in the cup of yogurt. Furthermore, we can say that the energy contribution from the protein content of the cup of yogurt is directly proportional to the specific energy content of protein and the mass of protein in the cup of yogurt.

f. The volume of salt solution left (the expression $41.8 - (2.1 \times 1.5 + 4.7 \times 2)$) is equal to the difference between the volume of salt solution I started with (the term 41.8) and the volume of salt solution I used (the term $(2.1 \times 1.5 + 4.7 \times 2)$). Furthermore, we can say that the volume of salt solution left is directly superpositional to the volume of salt solution I started with and inversely superpositional to the volume of salt solution that I used.

The volume of salt solution that I used (the term $(2.1 \times 1.5 + 4.7 \times 2)$) is equal to the sum of the volume of salt solution I used this morning (the term 2.1×1.5) and the volume of salt solution I used this afternoon (the term 4.7×2). Furthermore, we can say that the volume of salt solution that I used is directly superpositional to the volume of salt solution that I used this morning and the volume of salt solution that I used this afternoon.

The volume of salt solution that I used this morning (the term 2.1×1.5) is equal to the product of the rate of use of the salt solution this morning (the factor 2.1) and the duration of use of the salt solution this morning (the factor 1.5). Furthermore, we can say that the volume of salt solution that I used this morning is directly proportional to the rate of use of the salt solution this morning and the duration of use of the salt solution this morning.

The volume of salt solution that I used this afternoon (the term 4.7×2) is equal to the product of the rate of use of the salt solution this afternoon (the factor 4.7) and the duration of use of the salt solution this afternoon (the factor 2). Furthermore, we can say that the volume of salt solution that I used this afternoon is directly proportional to the rate of use of the salt solution this afternoon and the duration of use of the salt solution this afternoon.

g. The mass of 2.3 mol of C_3H_8 (the term $2.3\,(12.01 \times 3 + 1.008 \times 8)$) is equal to the product of the molar mass of C_3H_8 (the factor $(12.01 \times 3 + 1.008 \times 8)$) and the amount of C_3H_8 (the factor 2.3). Furthermore, we can say that the mass of 2.3 mol of $C_3\,H_8$ is directly proportional to the molar mass of C_3H_8 and the amount of C_3H_8.

The molar mass of C_3H_8 (the factor $(12.01 \times 3 + 1.008 \times 8)$) is equal to the mass of C atoms in 1 mol of C_3H_8 (the term 12.01×3) and the mass of H atoms in 1 mol of C_3H_8 (the term 1.008×8). Furthermore, we can say that the molar mass of C_3H_8 is directly superpositional to the mass of the C atoms in 1 mol of C_3H_8 and the mass of the H atoms in 1 mol of

C_3H_8.

The mass of C atoms in 1 mol of C_3H_8 (the term 12.01×3) is equal to the product of the molar mass of C (the factor 12.01) and the amount of C atoms in 1 mol of C_3H_8 (the factor 3). Furthermore, we can say that the mass of C atoms in 1 mol of C_3H_8 is directly proportional to the molar mass of C and the amount of C atoms in 1 mol of C_3H_8.

The mass of H atoms in 1 mol of C_3H_8 (the term 1.008×8) is equal to the product of the molar mass of H (the factor 1.008) and the amount of H atoms in 1 mol of C_3H_8 (the factor 8). Furthermore, we can say that the mass of H atoms in 1 mol of C_3H_8 is directly proportional to the molar mass of H and the amount of H atoms in 1 mol of C_3H_8.

h. The total distance that I covered today (the expression $110 \times 1.5 + 80 \times 2.5$) is equal to the sum of the distance I covered this morning (the term 110×1.5) and the distance I covered this afternoon (the term 80×2.5). Furthermore, we can say that the total distance that I covered today is directly superpositional to the distance that I covered this morning and the distance that I covered this afternoon.

The distance that I covered this morning (the term 110×1.5) is equal to the product of my speed this morning (the factor 110) and duration of travel this morning (the factor 1.5). Furthermore, we can say that the distance that I covered this morning is directly proportional to my speed this morning and the duration of travel this morning.

The distance that I covered this afternoon (the term 80×2.5) is equal to the product of my speed this afternoon (the factor 80) and duration of travel this afternoon (the factor 2.5). Furthermore, we can say that the distance that I covered this afternoon is directly proportional to my speed this afternoon and the duration of travel this afternoon.

i. The average mass of a box (the expression $\frac{11.9 \times 3 + 7.1 \times 2}{3+2}$) is equal to the quotient of the mass of all the boxes (the dividend $11.9 \times 3 + 7.1 \times 2$) and the number of boxes (the divisor $3+2$). Furthermore, we can say that the average mass of a box is directly proportional to the mass of all the boxes and inversely superpositional to the number of boxes.

The mass of all the boxes (the dividend $11.9 \times 3 + 7.1 \times 2$) is equal to the sum of the mass of the 3 boxes with identical masses (the term 11.9×3) and the mass of the 2 boxes with identical masses (the term 7.1×2). Furthermore, we can say that the mass of all the boxes is directly superpositional to the mass of the 3 boxes of identical mass, and the mass of the 2 boxes of identical mass.

The mass of the 3 boxes of identical mass (the term 11.9×3) is equal to the product of the mass of 1 such box (the factor 11.9) and the number of such boxes (the factor 3). Furthermore, we can say that the mass of these boxes is directly proportional to the mass of one such box and the number of such boxes.

The mass of the 2 boxes of identical mass (the term 7.1×2) is equal to the product of the mass of 1 such box (the factor 7.1) and the number of such boxes (the factor 2). Furthermore, we can say that the mass of these boxes is directly proportional to the mass of one such box and the number of such boxes.

The total number of boxes (the divisor $3 + 2$) is equal to the sum of the number of boxes of mass 11.9 g (the term 3) and the number of boxes of mass 7.1 g (the term 2). Furthermore, we can say that the total number of boxes is directly superpositional to the number of boxes of mass 11.9 g and the number of boxes of mass 7.1 g.

j. The average mass of an O atom in the sample (the expression $16 \times 0.973 + 17 \times 0.026 + 18 \times 0.001$) is equal to the sum of the contribution to the average mass by the ^{16}O isotopes (the term 16×0.973), the contribution to the average mass by the ^{17}O isotopes (the term 17×0.026), and the contribution to the average mass by the ^{18}O isotopes (the term 18×0.001). Furthermore, we can say that the average mass of an O atom in the sample is directly superpositional the contribution to the average mass by the ^{16}O isotopes, contribution to the average mass by the ^{17}O isotopes and contribution to the average mass by the ^{18}O isotopes.

The contribution to the average mass by the ^{16}O isotopes (the term 16×0.973) is equal to the product of the mass of an ^{16}O isotope (the factor 16) and the relative abundance of the ^{16}O isotope (the factor 0.973). Furthermore, we can say that the contribution to the average mass by the ^{16}O isotopes is directly proportional to the mass of an ^{16}O isotope and the relative abundance of the ^{16}O isotope.

The contribution to the average mass by the ^{17}O isotopes (the term 17×0.026) is equal to the product of the mass of an ^{17}O isotope (the factor 17) and the relative abundance of the ^{17}O isotope (the factor 0.026). Furthermore, we can say that the contribution to the average mass by the ^{17}O isotopes is directly proportional to the mass of an ^{17}O isotope and the relative abundance of the ^{17}O isotope.

The contribution to the average mass by the ^{18}O isotopes (the term 18×0.001) is equal to the product of the mass of an ^{18}O isotope (the factor 18) and the relative abundance of the ^{18}O isotope (the factor 0.001). Furthermore, we can say that the contribution to the average mass by the ^{18}O isotopes is directly proportional to the mass of an ^{18}O isotope and the relative abundance of the ^{18}O isotope.

k. My average speed today (the expression $\frac{110 \times 1.5 + 85 \times 3}{1.5 + 3}$) is equal to the quotient of distance that I covered today (the dividend $110 \times 1.5 + 85 \times 3$) and travel time today (the divisor $1.5 + 3$). Furthermore, we can say that my average speed today is directly proportional to the distance that I covered today and inversely proportional to the travel time today.

The distance that I covered today (the dividend $110 \times 1.5 + 85 \times 3$) is equal to the sum of distance that I covered this morning (the term 110×1.5) and distance that I covered this afternoon (the term 85×3). Furthermore, we can say that the distance that I covered today is directly superpositional to the distance that I covered this morning and the distance that I covered this afternoon.

The distance that I covered this morning (the term 110×1.5) is equal to the product of my speed this morning (the factor 110) and duration of travel this morning (the factor 1.5). Furthermore, we can say that the distance that I covered this morning is directly proportional to my speed this morning and the duration of travel this morning.

The distance that I covered this afternoon (the term 85×3) is equal to the product of my speed this afternoon (the factor 85) and duration of travel this afternoon (the factor 3). Furthermore, we can say that the distance I covered this afternoon is directly proportional to my speed this afternoon and the duration of travel this afternoon.

Total travel time today (the divisor $1.5 + 3$) is equal to the duration of travel this morning (the term 1.5) and duration of travel this afternoon (the term 3). Furthermore, we can say that the travel time today is directly superpositional to the duration of travel this morning and the duration of travel this afternoon.

Exercise Set 7

1. Formal reading of the phrase 0.25 *of* 20, i.e., *twenty-five hundredths of* 20, implies that 20 should be divided into 100 parts and that 25 parts should be taken. The mathematical expression 0.25×20 requires that we multiply 25 by 20, which in effect performs the second objective, and then place the decimal point, which in effect performs the first objective.[8] The expression,

---

[8]There is flexibility in the order in which multiplications and divisions are performed in an expression that consists of a chain of these operations. As an example, the sequence of steps *divide something into* 3 *parts and then take* 2 *parts by multiplying the result by* 2, a sequence that would map onto $\div 3 \times 2$, would result in a value that has the same size as the result of the sequence of steps *take* 2 *of those things, i.e., multiply the thing by* 2, *and then divide the result into* 3 *parts*, a sequence that would map onto $\times 2 \div 3$. To see this equivalence, represent the entity in question as a circle, divide the area that is covered by the circle into 3 parts and then shade 2 parts. Look at the total size of the shaded parts. Now, start a new drawing. Representing the entity in question as a circle, draw 2 circles and then divide the area that is covered by both circles into 3 parts. Shade one of the parts. This should have the same size as the shaded parts of the previous drawing.

This argument shows that the order in which multiplications and divisions in a chain are performed is immaterial. In the problem above, we can, therefore, switch order in the interpretation *divide* 20 *into* 100 *parts and then take* 25 *parts*, i.e., $20 \div 100 \times 25$, and instead follow the sequence in the phrase *multiply* 20 *by* 25 *and then divide into* 100 *parts*, i.e., $20 \times 25 \div 100$. The choice has tactical advantages: When working such problems out

0.25×20, then, performs the steps in the interpretation of the phrase 0.25 *of* 20.

We let the reader justify the equivalence between the phrase 25% *of* 20 and the mathematical expression $25\% \times 20$.

2. a.

m mass of the solution (g)

$$m = \frac{12.8}{4.5} \times 15.6$$

$$m = 44.4 \text{ g}$$

The solution has a mass of 44.4 g.

b.

V volume of the solution (L)

$$V = \frac{4.5}{12.8} \times 62.8$$

$$V = 22.1 \text{ L}$$

The solution has a volume of 22.1 L.

c.

n tablespoons of oil needed (1)

$$n = \frac{2.5}{1.5} \times 3.5$$

$$n = 5.8 \text{ tablespoons}$$

5.8 tablespoons of oil are needed.

d.

n cups of pop corn kernels (1)

$$n = \frac{1.5}{2.5} \times 1.75$$

$$n = 1.05 \text{ pop corn kernels}$$

1.05 cups of pop corn kernels should be used.

e.

V volume of fuel needed (L)

$$V = \frac{7.5}{80} \times 135$$

$$V = 12.66 \text{ L}$$

12.66 L of fuel will be needed.

---

by hand, it is easier to multiply 20 by 25, and then divide by 100 by placing the decimal point than dividing 20 by 100 and then multiplying the result by 25.

f.

d distance I can cover (km)

$$d = \frac{80}{7.5} \times 4.5$$

$d = 48$ km

I can cover 48 km.

g.

a extra amount of money Stephanie can make ($)

$$a = \frac{231}{16.5} \times 3.5$$

$a = \$49$

Stephanie can make an extra $49.

h.

t extra work time needed (h)

$$t = \frac{16.5}{231} \times 69$$

$t = 4.9$ h

Stephanie needs to work 4.9 h more.

i.

t time before I run out of fuel (h)

$$t = \frac{1}{120} \times \frac{100}{6.85} \times 15.75$$

$t = 1.9$ h

I will run out of fuel in 1.9 h.

j.

c cost of fuel for the trip ($)

$$c = \frac{1.136}{1} \times \frac{7.2}{100} \times 325$$

$c = \$26.58$

The cost of fuel for the trip is $26.58.

3. a.

t duration of the presentation (h)

$$t = \frac{1}{60} \times 135$$

$t = 2.25$ h

The presentation took 2.25 h.

b.

m mass of the newborn (lb)

$$m = \frac{2.2}{1} \times 3.2$$

$m = 7.04$ lb

The newborn has a mass of 7.04 lb.

c.

m mass of the newborn (lb)

$$m = \frac{1}{0.454} \times 3.2$$

$m = 7.05$ lb

The newborn has a mass of 7.05 lb.

d.

V volume of medication (oz)

$$V = \frac{0.034}{1} \times 1200$$

$V = 40.8$ oz

The medication has a volume of 40.8 oz.

e.

V volume of medication (oz)

$$V = \frac{1}{29.6} \times 1200$$

$V = 40.5$ oz

The medication has a volume of 40.5 oz.

f.

L length of the scar (mm)

$$L = \frac{1}{0.1} \times \frac{2.54}{1} \times 1.5$$

$L = 38.1$ mm

The scar has a length of 38.1 mm.

g.

m mass of the liquid (lb)

$$m = \frac{0.0022}{1} \times \frac{1}{0.001} \times 3.8$$

$m = 8.36$ lb

The liquid has a mass of 8.36 lb.

h.

a amount of money I have (€)

$$a = \frac{0.716\,29}{1} \times \frac{0.974\,44}{1} \times 250$$

$a = 174.50$ €

I have 174.50 €.

i.

a amount of money I have (€)

$$a = \frac{1}{1.395\,09} \times \frac{1}{1.025\,23} \times 250$$

$a = 174.79$ €

I have 174.79 €.

j.

p the pressure (kPa)

$$p = \frac{0.001}{1} \times \frac{101\,325}{1} \times 1.2$$

$p = 121.59$ kPa

The pressure is 121.59 kPa.

k.

d distance from New York City to Los Angeles (km)

$$d = \frac{1}{1000} \times \frac{1}{1.09} \times \frac{1760}{1} \times 2780$$

$d = 4489$ km

The distance from New York City to Los Angeles is 4489 km.

l.

E heat generated by the reaction (Cal)

$$E = \frac{0.001}{1} \times \frac{1}{4.184} \times \frac{1000}{1} \times 3500$$

$E = 836.5$ Cal

The reaction generated 836.5 Cal of heat.

4. a.

m mass of $C_6H_{12}O_6$ (g)

$$m = \frac{180.156}{1} \times 4.8$$

$m = 864.7$ g

The $C_6H_{12}O_6$ has a mass of 864.7 g.

b.

n amount of $C_6H_{12}O_6$ (mol)

$n = \dfrac{1}{180.156} \times 76$

$n = 0.422$ mol

There is 0.422 mol of $C_6H_{12}O_6$.

c.

m mass of O_2 (g)

$m = \dfrac{32}{1} \times 1900$

$m = 60\,800$ g

The O_2 has a mass of 60 800 g.

d.

n amount of O_2 (mol)

$n = \dfrac{1}{32} \times 1200$

$n = 37.5$ mol

37.5 mol of O_2 corresponds to the given mass of O_2.

e.

m mass of NO_3 (amu)

$m = \dfrac{62.01}{1} \times 20$

$m = 1240$ amu

The NO_3 has a mass of 1240 amu.

f.

n amount of SO_2 (1)

$n = \dfrac{1}{64.07} \times 80\,087.5$

$n = 1250$ molecules

There are 1250 molecules of SO_2.

5. a.

n amount of CO_2 generated (mol)

$n = \dfrac{6}{2} \times 4.5$

$n = 13.5$ mol

13.5 mol of CO_2 were generated by the reaction.

b.

n amount of Al used (mol)

$$n = \frac{4}{2} \times 520$$

$n = 1040$ mol

1040 mol of Al were used by the reaction.

c.

n amount of H^+ generated (mol)

$$n = \frac{2}{1} \times 120$$

$n = 240$ mol

240 mol of H^+ were generated by the dissociation.

d.

n amount of H_2O generated (mol)

$$n = \frac{2}{518} \times 1420$$

$n = 5.48$ mol

5.48 mol of H_2O were generated by the reaction.

e.

n amount of O_3 generated (mol)

$$n = \frac{1}{1} \times 125$$

$n = 125$ mol

125 mol of O_3 can be generated.

f.

n amount of Mg needed (mol)

$$n = \frac{3}{2} \times 42.5$$

$n = 63.75$ mol

63.75 mol of Mg is needed.

g.

E heat taken up by the reaction (Cal)

$$E = \frac{44}{2} \times 32.8$$

$E = 721.6$ Cal

721.6 Cal of heat is taken up by the reaction.

6. a.   i.

n amount of CO_2 generated (mol)

$$n = \frac{4}{2} \times \frac{1}{30.068} \times 82.7$$

$n = 5.5$ mol

5.5 mol of CO_2 were generated by the reaction.

ii.

m mass of C_2H_6 used (g)

$$m = \frac{30.068}{1} \times \frac{2}{5} \times 122$$

$m = 1467$ g

1467 g of C_2H_6 were used by the reaction.

iii.

n amount of H_2O generated (mol)

$$n = \frac{2}{4} \times \frac{1}{44.01} \times 42.8$$

$n = 0.486$ mol

0.486 mol of H_2O were generated by the reaction.

b.   i.

m mass of CO_2 generated (g)

$$m = \frac{44.01}{1} \times \frac{2}{3} \times \frac{1}{32.00} \times 154.7$$

$m = 141.8$ g

141.8 g of CO_2 were generated by the reaction.

ii.

m mass of H_2O generated (g)

$$m = \frac{18.016}{1} \times \frac{3}{1} \times \frac{1}{46.068} \times 1800$$

$m = 2112$ g

2112 g of H_2O were generated by the reaction.

iii.

m mass of O_2 used (g)

$$m = \frac{32.00}{1} \times \frac{3}{1} \times \frac{1}{46.068} \times 120$$

$m = 250.1$ g

250.1 g of O_2 were used by the reaction.

c.   i.

m mass of CO generated (g)

$$m = \frac{28.01}{1} \times \frac{4}{2511} \times 12\,000$$

$m = 535.4$ g

535.4 g of CO were generated by the reaction.

ii.

E heat generated (kJ)

$$E = \frac{2511}{2} \times \frac{1}{26.036} \times 450$$

$E = 21\,700$ kJ

21 700 kJ of heat were generated by the reaction.

d.   i.

m mass of H_2O used (g)

$$m = \frac{18.016}{1} \times \frac{2}{518} \times 38\,500$$

$m = 2678$ g

2678 g of H_2O were used by the reaction.

ii.

E heat taken up (kJ)

$$E = \frac{518}{2} \times \frac{1}{2.016} \times 72.6$$

$E = 9327$ kJ

9327 kJ of heat were taken up by the reaction.

7. a.

a amount of discount ($)
$a = 0.15 \times 32.99$
$a = \$4.95$
The amount of discount is $4.95.

b.

a amount of discount ($)
$a = 0.1 \times 125.00$
$a = \$12.50$
The amount of discount is $12.50.

c.

p_s sale price ($)
$p_s = (1 - 0.1)\,(25.50)$
$p_s = 0.9 \times 25.50$
$p_s = \$22.95$
The sale price is $22.95.

d.

p_s sale price (\$)
$p_s = (1 - 0.125)(87.00)$
$p_s = 0.875 \times 87.00$
$p_s = \$76.13$
The sale price is \$76.13.

e.

a amount of tax (\$)
$a = 0.13 \times 120$
$a = \$15.60$
The amount of tax is \$15.60.

f.

a amount of tax (\$)
$a = 0.125 \times 67.25$
$a = \$8.41$
The amount of tax is \$8.41.

g.

t grand total (\$)
$t = (1 + 0.13)(52.80)$
$t = 1.13 \times 52.80$
$t = \$59.66$
The grand total is \$59.66.

h.

t grand total (\$)
$t = (1 + 0.08)(520.00)$
$t = 1.08 \times 520.00$
$t = \$561.60$
The grand total is \$561.60.

i.

a amount of tax (\$)
$a = 0.15(1 - 0.05)(25.99)$
$a = 0.15 \times 0.95 \times 25.99$
$a = \$3.70$
The amount of tax is \$3.70.

j.

a amount of tax (\$)
$a = 0.13(1 - 0.125)(5200)$
$a = 0.13 \times 0.875 \times 5200$
$a = \$591.50$
The amount of tax is \$591.50.

k.

t grand total (\$)
$t = (1 + 0.07)(1 - 0.1)(14.80)$
$t = 1.07 \times 0.9 \times 14.80$
$t = \$14.25$
The grand total is \$14.25.

l.

t grand total ($)
$t = (1 + 0.13)(1 - 0.15)(550)$
$t = 1.13 \times 0.85 \times 550$
$t = \$528.28$
The grand total is $528.28.

m.

a amount of tip ($)
$a = 0.15 \times 37.50$
$a = \$5.63$
The amount of tip is $5.63.

n.

t total amount to pay ($)
$t = (1 + 0.15)(22.20)$
$t = 1.15 \times 22.20$
$t = \$25.53$
I should pay $25.53 in total.

8. a.

n number of graduates who are expected to find employment within a year (1)
$n = 0.823 \times 320$
$n = 263$
263 graduates are expected to find employment within a year.

b.

n number of eligible voters who voted (1)
$n = 0.628 \times 12\,430$
$n = 7806$
7806 eligible voters voted.

c.

n number of polled individuals who did not pick Part A as their favourite party (1)
$n = (1 - 0.428)(1200)$
$n = 0.572 \times 1200$
$n = 686$
686 individuals who were polled did not pick Part A as their favourite party.

d.

n this year's moose population in the park (1)
$n = (1 - 0.085)(1820)$
$n = 0.915 \times 1820$
$n = 1665$
This year the park has a moose population of 1665.

e.

n number of units produced this year (1)
$n = (1 + 0.112)(13\,000)$
$n = 1.112 \times 13\,000$
$n = 14\,456$
14 456 units were produced this year.

f.

n enrollment in the program this year (1)
$n = (1 + 0.079)(13\,00)$
$n = 1.079 \times 1300$
$n = 1403$
This year the enrollment is 1403.

g.

n number of participants who will experience side effects
and will require hospitalization (1)
$n = 0.038 \times 0.245 \times 400$
$n = 4$
4 participants are expected to experience side effects and
require hospitalization.

h.

n number of club members who left the club, citing the
deterioration of services as the reason for leaving (1)
$n = 0.63 \times 0.223 \times 1250$
$n = 176$
176 club members who left the club cited the deterioration
of services as the reason for leaving.

i.

n number of fish that did not die and reproduced (1)
$n = 0.619\,(1 - 0.428)\,(26\,300)$
$n = 0.619 \times 0.572 \times 26\,300$
$n = 9312$
9312 fish did not die and reproduced.

j.

n number of those who did not vote in favour of amending
the constitution and identified themselves as supporters
of the ABC political party (1)
$n = 0.893\,(1 - 0.725)\,(732\,300)$
$n = 0.893 \times 0.275 \times 732\,300$
$n = 179\,835$
179 835 voters who did not vote in favour of amending the
constitution identified themselves as supporters
of the ABC political party.

k.

n number of part time students taking fewer than
three courses (1)
$n = (1 - 0.785)\,(0.828)\,(120)$
$n = 0.215 \times 0.828 \times 120$
$n = 21$
It is expected that 21 part time students in the class
will be taking fewer than three courses.

l.

n number of respondents who rated the subway service as unsatisfactory citing a reason other than unreliability as the reason for their choice (1)

$n = (1 - 0.252)(0.754)(1020)$

$n = 0.748 \times 0.754 \times 1020$

$n = 575$

575 respondents rated the subway service as unsatisfactory citing a reason other than unreliability as the reason for their choice.

m.

n number of eligible voters who voted and did not reject the motion (1)

$n = (1 - 0.612)(1 - 0.312)(370)$

$n = 0.388 \times 0.688 \times 370$

$n = 99$

99 eligible voters who voted did not reject the motion.

n.

n number of people who are unemployed but are expected not to find employment in the next six months (1)

$n = (1 - 0.42)(1 - 0.928)(370\,000)$

$n = 0.58 \times 0.072 \times 370\,000$

$n = 15\,451$

15 451 unemployed individuals are expected not to find employment in the next six months.

9. a.

r conversion factor for converting the unit of mass from g to lb (lb/g)

$$r = \frac{1}{0.4536} \times \frac{1}{1000}$$

$r = 0.002\,205 \text{ lb/g}$

The conversion factor for converting the unit of mass from g to lb is 0.002 205 lb/g.

b.

r conversion factor for converting the unit of time from s to h (h/s)

$$r = \frac{1}{60} \times \frac{1}{60}$$

$r = 0.000\,278 \text{ h/s}$

The conversion factor for converting the unit of time from s to h is 0.000 278 h/s.

c.

r conversion factor for converting the unit of pressure from kPa to atm (atm/kPa)

$$r = \frac{1}{760.0} \times \frac{1}{133.322} \times \frac{1}{0.001}$$

$r = 0.009\,87$ atm/kPa

The conversion factor for converting the unit of pressure from kPa to atm is $0.009\,87$ atm/kPa.

d.

r percentage of the population who got sick and were hospitalized (1)

$r = 0.02 \times 0.19$

$r = 0.0038$

0.38% of the population got sick and were hospitalized.

e.

r percentage of the students who graduate with a GPA of at least 2.00 (1)

$r = (1 - 0.213)(1 - 0.173)$

$r = 0.787 \times 0.827$

$r = 0.651$

65.1% of the students graduate with a GPA of at least 2.00.

f.

r fraction of the budget given to the marketing department (1)

$$r = \frac{1}{4}\left(1 - \frac{1}{8}\right)$$

$$r = \frac{1}{4} \times \frac{8 - 1}{8}$$

$$r = \frac{1}{4} \times \frac{7}{8}$$

$$r = \frac{7}{32}$$

$\frac{7}{32}$ of the budget is given to the marketing department.

10. We can rewrite $\frac{q_3 q_2 q_1}{q_4 q_5 q_6}$ as $\frac{q_3 q_2}{q_5} \times \frac{q_1}{q_4 q_6}$, i.e.,

$$\frac{q_3 q_2 q_1}{q_4 q_5 q_6} = \frac{q_3 q_2}{q_5} \times \frac{q_1}{q_4 q_6}$$

Since the right side consists of a single term and $\frac{q_1}{q_4 q_6}$ is a factor within this term, the expression $\frac{q_3 q_2 q_1}{q_4 q_5 q_6}$ is directly proportional to $\frac{q_1}{q_4 q_6}$.

11. a. q_1, q_2, q_3, $q_2 q_1$, $q_3 q_1$, $q_3 q_2$, $q_3 q_2 q_1$

b. q_1, q_2, $\frac{1}{q_3}$, $q_2 q_1$, $\frac{q_1}{q_3}$, $\frac{q_2}{q_3}$, $\frac{q_2 q_1}{q_3}$

c. q_1, $\frac{1}{q_2}$, $\frac{1}{q_3}$, $\frac{q_1}{q_2}$, $\frac{q_1}{q_3}$, $\frac{1}{q_2 q_3}$, $\frac{q_1}{q_2 q_3}$

d. $\frac{1}{q_1}$, $\frac{1}{q_2}$, $\frac{1}{q_3}$, $\frac{1}{q_1 q_2}$, $\frac{1}{q_1 q_3}$, $\frac{1}{q_2 q_3}$, $\frac{1}{q_1 q_2 q_3}$

12. It means that, if q_2 is directly proportional to q_1, then q_1 is directly proportional to q_2.[9]

Exercise Set 8

1. a.

t time it takes to assemble the speakers (day)

$$t = \frac{12 \times 18}{15}$$

$t = 14.4$ days

It will take 14.4 days to assemble the speakers.

b.

t time it takes the cleaners to clean the office (h)

$$t = \frac{8.5 \times 2}{6}$$

$t = 2.83$ h

It will take 2.83 h.

c.

v speed for the given travel time (km/h)

$$v = \frac{80 \times 4.5}{4}$$

$v = 90$ km/h

I will have to drive at a speed of 90 km/h.

d.

t time it takes to get to Calgary (h)

$$t = \frac{110 \times 1.8}{90}$$

$t = 2.2$ h

It will take 2.2 h to get to Calgary.

---

[9]The assertion q_2 *is directly proportional to* q_1 tells us that, if we were to change q_1 by doubling it, tripling it, halving it, etc., then q_2 would also double or triple or reduce to half its value. The assertion that q_1 *is directly proportional to* q_2 states that, if we were to change q_2 by doubling it, tripling it, halving it, etc., then q_1 would also double or triple or reduce to half its value. Symmetry asserts that if one of these holds, then the other holds as well.

e.

p pressure of air (Pa)

$$p = \frac{340\,000 \times 0.75}{0.62}$$

$p = 411\,290$ Pa

The air will have a pressure of 411 290 Pa.

f.

V volume of the tank (m^3)

$$V = \frac{3.95 \times 428\,000}{380\,000}$$

$V = 4.45$ m^3

The helium will have a volume of 4.45 m^3.

g.

I current through the wire (A)

$$I = \frac{3.5 \times 2.5}{4.1}$$

$I = 2.1$ A

A current of 2.1 A will flow through the wire.

h.

R wire's resistance (Ω)

$$R = \frac{3.5 \times 2.5}{7.8}$$

$R = 1.1\ \Omega$

The wire has a resistance of 1.1 Ω.

i.

a object's acceleration (m/s^2)

$$a = \frac{20.9 \times 2.45}{32.8}$$

$a = 1.56$ m/s^2

The object will experience an acceleration of 1.56 m/s^2.

j.

m mass of object (kg)

$$m = \frac{12.8 \times 1.53}{1.73}$$

$m = 11.3$ kg

The object has a mass of 11.3 kg.

2. We can rewrite $\frac{q_3 q_2 q_1}{q_4 q_5 q_6}$ as $\frac{q_3 q_2}{q_5 \left(\frac{q_4 q_6}{q_1} \right)}$, i.e.,

$$\frac{q_3 q_2 q_1}{q_4 q_5 q_6} = \frac{q_3 q_2}{q_5 \left(\frac{q_4 q_6}{q_1} \right)}$$

Since the right side consists of a single term and $\frac{q_4 q_6}{q_1}$ is a factor in the divisor of that term, the expression $\frac{q_3 q_2 q_1}{q_4 q_5 q_6}$ is inversely proportional to $\frac{q_4 q_6}{q_1}$.

3. a. $q_1, q_2, q_3, q_1 q_2, q_1 q_3, q_2 q_3, q_1 q_2 q_3$

   b. $\frac{1}{q_1}, \frac{1}{q_2}, q_3, \frac{1}{q_1 q_2}, \frac{q_3}{q_1}, \frac{q_3}{q_2}, \frac{q_3}{q_1 q_2}$

   c. $\frac{1}{q_1}, q_2, q_3, \frac{q_2}{q_1}, \frac{q_3}{q_1}, q_2 q_3, \frac{q_2 q_3}{q_1}$

   d. $\frac{1}{q_1}, \frac{1}{q_2}, \frac{1}{q_3}, \frac{1}{q_1 q_2}, \frac{1}{q_1 q_3}, \frac{1}{q_2 q_3}, \frac{1}{q_1 q_2 q_3}$

4. It means that, if q_2 is inversely proportional to q_1, then q_1 is inversely proportional to q_2.[10]

Exercise Set 9.2

1.

m mass of protein in the glass of milk (g)

$79.4 = 4 \times 12.5 + 9 \times 1.4 + 4 \times m$

2.

e specific energy content of protein (Cal/g)

$164.6 = 4 \times 16 + 9 \times 4.2 + e \times 15.7$

3.

E energy content of the cup of tea (Cal)

$-85 = 37 \times 3 + 52 \times 2 + E - 80 \times 2 - 320 \times 0.5$

4.

c cost of a notebook ($/notebook)

$54.92 = 2.50 \times 6 + c \times 8$

5.

n number of rolls of string purchased (1)

$13.80 = 1.20 \times 6.5 + 2.00 \times n$

---

[10]The assertion q_2 *is inversely proportional to* q_1 tells us that, if we were to change q_1 by doubling it, tripling it, halving it, etc., then q_2 would reduce to half its value, reduce to a third of its value, or double, etc. The assertion that q_1 *is inversely proportional to* q_2 states that, if we were to change q_2 by doubling it, tripling it, halving it, etc., then q_1 would reduce to half its value, reduce to a third of its value, or double. Symmetry asserts that if one of these holds, then the other holds as well.

6.

    r number of packages of printer paper per box (packages/box)

    $1076.40 = 2.99 \times r \times 20$

7.

    c cost of a box of masks (\$/box)

    $960 = c \times 12 \times 25$

8.

    c cost of a package of printer paper (\$/package)

    $538.80 = 1.50 \times 200 + c \times 8 \times 15$

9.

    a amount of money I started with (\$)

    $35.01 = a - 16.99 - 2.00 \times 3$

10.

    V volume of medication I started with (mL)

    $332.2 = V - 5.65 \times 12$

11.

    t duration of use of medication (min)

    $108 = 180 - 2.4 \times t$

12.

    r rate of use of medication (mL/min)

    $108 = 180 - r \times 30$

13.

    M atomic mass of the unknown atom, X (amu/atom)

    $30.068 = 12.01 \times 2 + M \times 6$

14.

    M molar mass of the unknown atom, X (g/mol)

    $30.068 = M \times 2 + 1.008 \times 6$

15.

    n number of C atoms in a molecule of C_nH_6 (1)

    $30.068 = 12.01 \times n + 1.008 \times 6$

16.

    M atomic mass of the unknown atom, X (amu/atom)

    $2690.94 = (32.07 + M \times 2) \times 42$

17.

M molar mass of the unknown atom, X (g/mol)

$2690.94 = (M + 16.00 \times 2) \times 42$

18.

M atomic mass of the unknown atom, X (amu/atom)

$74.096 = 40.08 + (M + 1.008) \times 2$

19.

M molar mass of the unknown atom, X (g/mol)

$253.922 = M \times 2 + (12.01 + 16.00 \times 3) \times 4$

20.

M molar mass of the unknown atom, X (g/mol)

$1808.976 = \left[22.99 \times 4 + (M + 16.00 \times 4) \times 3 \right] \times 4.8$

21.

n number of molecules of $(NH_4)_3PO_4$ (1)

$15\,058.696 = \left[(14.01 + 1.008 \times 4) \times 3 + 30.97 + 16.00 \times 4 \right] \times n$

22.

t temperature on the third day (°C)

$15.8 = \dfrac{12 + 15 + t + 14 + 20}{5}$

23.

a maximum amount of money Marie can spend ($)

$2000 = \dfrac{2100 + 1600 + a}{3}$

24.

c cost of a Type C computer system ($/system)

$1043.33 = \dfrac{1200 \times 8 + 820 + c \times 3}{8 + 1 + 3}$

25.

r rate of pay at my third job ($/h)

$15.04 = \dfrac{14 \times 15 + 16 \times 10 + r \times 2}{15 + 10 + 2}$

26.

r training time per week per worker in Section C (h/week/worker)

$4 = \dfrac{3 \times 120 + 5 \times 250 + r \times 320}{120 + 250 + 320}$

27.

g grade point in CHEM101 for a GPA of 3.00 (1)

$$3.00 = \frac{3.33 \times 3 + 1.67 \times 5 + g \times 4}{3 + 5 + 4}$$

28.

g grade point in COMM1020 for a GPA of 3.40 (1)

$$3.40 = \frac{4.00 \times 3 + 2.67 \times 2 + 2.33 \times 4 + g \times 3}{3 + 2 + 4 + 3}$$

29.

p price of the hat ($)

$4.50 = 0.2 \times p$

30.

r rate of discount (1)

$16.38 = (1 - r) \times 18.20$

31.

p price of the history textbook ($)

$3.38 = 0.13 \times p$

32.

p price of the gloves ($)

$17.48 = (1 + 0.13) \times (1 - 0.15) \times p$

33.

r rate of tax (1)

$17.48 = (1 + r) \times (1 - 0.15) \times 18.20$

34.

p price of the laptop ($)

$132.60 = 0.13 \times (1 - 0.15) \times p$

35.

r rate of discount (1)

$2.90 = 0.13 \times (1 - r) \times 24.80$

36.

r rate of tip (1)

$3.00 = r \times 19.10$

37.

n number of respondents (1)

$1246 = 0.82 \times n$

38.

n number of respondents (1)

$274 = (1 - 0.82) \times n$

39.

r percent rise in population (1)

$5544 = r \times 132\,000$

40.

n population of the town ten years ago (1)

$5544 = 0.042 \times n$

41.

n number of eligible voters (1)

$76 = 0.18 \times 0.78 \times n$

42.

n number of eligible voters (1)

$346 = (1 - 0.18) \times 0.78 \times n$

43.

r percentage of eligible voters who voted (1)

$76 = 0.18 \times r \times 541$

44.

r percentage of eligible voters who voted (1)

$346 = (1 - 0.18) \times r \times 541$

45.

r percentage of eligible voters who voted and opposed the proposed changes (1)

$76 = r \times 0.78 \times 541$

46.

r percentage of eligible voters who voted and opposed the proposed changes (1)

$346 = (1 - r) \times 0.78 \times 541$

47.

n number of frogs in the sample (1)

$62 = 0.15 \times 0.34 \times n$

48.

r percentage of participants who experienced side effects and needed to be hospitalized (1)

$70 = (1 - r) \times 0.61 \times 120$

49.

r rate of decrease in population of Blue Lake City in the nineties (1)

$181\,440 = (1 - r) \times (1 + 0.12) \times 180\,000$

50.

n number of people who commute to work daily (1)

$45\,000 = (1 - 0.72) \times (1 - 0.33) \times n$

Exercise Set 9.3A

1. a.

m mass of protein in the glass of milk (g)

$79.4 = 4 \times 12.5 + 9 \times 1.4 + 4 \times m$

$4 \times 12.5 + 9 \times 1.4 + 4 \times m = 79.4$

$4 \times m = 79.4 - 4 \times 12.5 - 9 \times 1.4$

$m = \dfrac{79.4 - 4 \times 12.5 - 9 \times 1.4}{4}$

$m = \dfrac{79.4 - 50 - 12.6}{4}$

$m = \dfrac{16.8}{4}$

$m = 4.2$

The glass of milk contains 4.2 g of protein.

b.

m mass of protein in the glass of milk (g)

$79.4 = 4 \times 12.5 + 9 \times 1.4 + 4m$

$4 \times 12.5 + 9 \times 1.4 + 4m = 79.4$

$50 + 12.6 + 4m = 79.4$

$62.6 + 4m = 79.4$

$4m = 79.4 - 62.6$

$4m = 16.8$

$m = \dfrac{16.8}{4}$

$m = 4.2$

The glass of milk contains 4.2 g of protein.

2. a.

e specific energy content of protein (Cal/g)

$164.6 = 4 \times 16 + 9 \times 4.2 + e \times 15.7$

$4 \times 16 + 9 \times 4.2 + e \times 15.7 = 164.6$

$e \times 15.7 = 164.6 - 4 \times 16 - 9 \times 4.2$

$e = \dfrac{164.6 - 4 \times 16 - 9 \times 4.2}{15.7}$

$e = \dfrac{164.6 - 64 - 37.8}{15.7}$

$e = \dfrac{62.8}{15.7}$

$e = 4$

The specific energy content of protein is 4 Cal/g.

b.

e specific energy content of protein (Cal/g)

$164.6 = 4 \times 16 + 9 \times 4.2 + 15.7e$

$4 \times 16 + 9 \times 4.2 + 15.7e = 164.6$

$64 + 37.8 + 15.7e = 164.6$

$101.8 + 15.7e = 164.6$

$15.7e = 164.6 - 101.8$

$15.7e = 62.8$

$e = \dfrac{62.8}{15.7}$

$e = 4$

The specific energy content of protein is 4 Cal/g.

3. a.

E energy content of the cup of tea (Cal)

$-85 = 37 \times 3 + 52 \times 2 + E - 80 \times 2 - 320 \times 0.5$

$37 \times 3 + 52 \times 2 + E - 80 \times 2 - 320 \times 0.5 = -85$

$E = -85 - 37 \times 3 - 52 \times 2 + 80 \times 2 + 320 \times 0.5$

$E = -85 - 111 - 104 + 160 + 160$

$E = 20$ Cal

The cup of tea contained 20 Cal of energy.

b.

E energy content of the cup of tea (Cal)

$-85 = 37 \times 3 + 52 \times 2 + E - 80 \times 2 - 320 \times 0.5$

$37 \times 3 + 52 \times 2 + E - 80 \times 2 - 320 \times 0.5 = -85$

$111 + 104 + E - 160 - 160 = -85$

$-105 + E = -85$

$E = -85 + 105$

$E = 20$ Cal

The cup of tea contained 20 Cal of energy.

4. a.

c cost of a notebook (\$/notebook)

$54.92 = 2.50 \times 6 + c \times 8$

$2.50 \times 6 + c \times 8 = 54.92$

$c \times 8 = 54.92 - 2.50 \times 6$

$c = \dfrac{54.92 - 2.50 \times 6}{8}$

$c = \dfrac{54.92 - 15}{8}$

$c = \dfrac{39.92}{8}$

$c = \$4.99$/notebook

The notebook cost \$4.99.

b.

c cost of a notebook (\$/notebook)

$54.92 = 2.50 \times 6 + 8c$

$2.50 \times 6 + 8c = 54.92$

$15 + 8c = 54.92$

$8c = 54.92 - 15$

$8c = 39.92$

$c = \dfrac{39.92}{8}$

$c = \$4.99/\text{notebook}$

The notebook cost \$4.99.

5. a.

n number of rolls of string purchased (1)

$13.80 = 1.20 \times 6.5 + 2.00 \times n$

$1.20 \times 6.5 + 2.00 \times n = 13.80$

$2.00 \times n = 13.80 - 1.20 \times 6.5$

$n = \dfrac{13.80 - 1.20 \times 6.5}{2.00}$

$n = \dfrac{13.80 - 7.80}{2.00}$

$n = \dfrac{6.00}{2.00}$

$n = 3$

I purchased 3 rolls of string.

b.

n number of rolls of string purchased (1)

$13.80 = 1.20 \times 6.5 + 2.00n$

$1.20 \times 6.5 + 2.00n = 13.80$

$7.80 + 2.00n = 13.80$

$2.00n = 13.80 - 7.80$

$2.00n = 6.00$

$n = \dfrac{6.00}{2.00}$

$n = 3$

I purchased 3 rolls of string.

6. a.

r number of packages of printer paper per box (packages/box)

$1076.40 \ = \ 2.99 \times r \times 20$

$2.99 \times r \times 20 \ = \ 1076.40$

$r \ = \ \dfrac{1076.40}{2.99 \times 20}$

$r \ = \ 18$

There are 18 packages of printer paper per box.

b.

r number of packages of printer paper per box (packages/box)

$1076.40 \ = \ 2.99\,(20)\,r$

$2.99\,(20)\,r \ = \ 1076.40$

$59.8r \ = \ 1076.40$

$r \ = \ \dfrac{1076.40}{59.8}$

$r \ = \ 18 \text{ packages/box}$

There are 18 packages of printer paper per box.

7. a.

c cost of a box of masks (\$/box)

$960 \ = \ c \times 12 \times 25$

$c \times 12 \times 25 \ = \ 960$

$c \ = \ \dfrac{960}{12 \times 25}$

$c \ = \ \$3.20/\text{box}$

A box of nasks costs \$3.20.

b.

c cost of a box of masks (\$/box)

$960 \ = \ 12\,(25)\,c$

$12\,(25)\,c \ = \ 960$

$300c \ = \ 960$

$c \ = \ \dfrac{960}{300}$

$c \ = \ \$3.20/\text{box}$

A box of nasks costs \$3.20.

8. a.

c cost of a package of printer paper (\$/package)

$538.80 = 1.50 \times 200 + c \times 8 \times 15$

$1.50 \times 200 + c \times 8 \times 15 = 538.80$

$c \times 8 \times 15 = 538.80 - 1.50 \times 200$

$c = \dfrac{538.80 - 1.50 \times 200}{8 \times 15}$

$c = \dfrac{538.80 - 300}{120}$

$c = \dfrac{238.80}{120}$

$c = \$1.99/\text{package}$

A package of printer paper costs \$1.99.

b.

c cost of a package of printer paper (\$/package)

$538.80 = 1.50\,(200) + 8\,(15)\,c$

$1.50\,(200) + 8\,(15)\,c = 538.80$

$300 + 120c = 538.80$

$120c = 538.80 - 300$

$120c = 238.80$

$c = \dfrac{238.80}{120}$

$c = \$1.99/\text{package}$

A package of printer paper costs \$1.99.

9. a.

a amount of money I started with (\$)

$35.01 = a - 16.99 - 2.00 \times 3$

$a - 16.99 - 2.00 \times 3 = 35.01$

$a = 35.01 + 16.99 + 2.00 \times 3$

$a = 35.01 + 16.99 + 6.00$

$a = \$58.00$

I started with \$58.00.

b.

a amount of money I started with ($)

$35.01 = a - 16.99 - 2.00 \times 3$

$a - 16.99 - 2.00 \times 3 = 35.01$

$a - 16.99 - 6.00 = 35.01$

$a - 22.99 = 35.01$

$a = 35.01 + 22.99$

$a = 58.00$

I started with $58.00.

10. a.

V volume of medication I started with (mL)

$332.2 = V - 5.65 \times 12$

$V - 5.65 \times 12 = 332.2$

$V = 332.2 + 5.65 \times 12$

$V = 332.2 + 67.8$

$V = 400$ mL

I started with 400 mL of medication.

b.

V volume of medication I started with (mL)

$332.2 = V - 5.65 \times 12$

$V - 5.65 \times 12 = 332.2$

$V - 67.8 = 332.2$

$V = 332.2 + 67.8$

$V = 400$ mL

I started with 400 mL of medication.

11. a.

t duration of use of medication (min)

$108 = 180 - 2.4 \times t$

$180 - 2.4 \times t = 108$

$-2.4 \times t = -180 + 108$

$t = \dfrac{-180 + 108}{-2.4}$

$t = \dfrac{180 - 108}{2.4}$

$t = \dfrac{72}{2.4}$

$t = 30$ min

I used the medication for 30 min.

b.

t duration of use of medication (min)

$108 = 180 - 2.4t$

$180 - 2.4t = 108$

$-2.4t = 108 - 180$

$-2.4t = -72$

$t = \dfrac{-72}{-2.4}$

$t = 30$ min

I used the medication for 30 min.

12. a.

r rate of use of medication (mL/min)

$108 = 180 - r \times 30$

$180 - r \times 30 = 108$

$-r \times 30 = -180 + 108$

$r = \dfrac{-180 + 108}{-30}$

$r = \dfrac{180 - 108}{30}$

$r = \dfrac{72}{30}$

$r = 2.4$ mL/min

I used the medication at a rate of 2.4 mL/min.

b.

r rate of use of medication (mL/min)

$108 = 180 - 30r$

$180 - 30r = 108$

$-30r = 108 - 180$

$-30r = -72$

$r = \dfrac{-72}{-30}$

$r = 2.4$ mL/min

I used the medication at a rate of 2.4 mL/min.

13. a.

M atomic mass of the unknown atom, X (amu/atom)

$30.068 = 12.01 \times 2 + M \times 6$

$12.01 \times 2 + M \times 6 = 30.068$

$M \times 6 = 30.068 - 12.01 \times 2$

$M = \dfrac{30.068 - 12.01 \times 2}{6}$

$M = \dfrac{30.068 - 24.02}{6}$

$M = \dfrac{6.048}{6}$

$M = 1.008$ amu/atom

The atomic mass of X is 1.008 amu/atom which identifies the atom as hydrogen.

b.

M atomic mass of the unknown atom, X (amu/atom)

$30.068 = 12.01 \times 2 + 6M$

$12.01 \times 2 + 6M = 30.068$

$24.02 + 6M = 30.068$

$6M = 30.068 - 24.02$

$6M = 6.048$

$M = \dfrac{6.048}{6}$

$M = 1.008$ amu/atom

The atomic mass of X is 1.008 amu/atom which identifies the atom as hydrogen.

14. a.

M molar mass of the unknown atom, X (g/mol)

$30.068 = M \times 2 + 1.008 \times 6$

$M \times 2 + 1.008 \times 6 = 30.068$

$M \times 2 = 30.068 - 1.008 \times 6$

$M = \dfrac{30.068 - 1.008 \times 6}{2}$

$M = \dfrac{30.068 - 6.048}{2}$

$M = \dfrac{24.02}{2}$

$M = 12.01$ g/mol

The molar mass of X is 12.01 g/mol which identifies the atom as carbon.

b.

M molar mass of the unknown atom, X (g/mol)

$30.068 = 2M + 1.008 \times 6$

$2M + 1.008 \times 6 = 30.068$

$2M + 6.048 = 30.068$

$2M = 30.068 - 6.048$

$2M = 24.02$

$M = \dfrac{24.02}{2}$

$M = 12.01$ g/mol

The molar mass of X is 12.01 g/mol which identifies the atom as carbon.

15. a.

n number of C atoms in a molecule of C_nH_6 (1)

$30.068 = 12.01 \times n + 1.008 \times 6$

$12.01 \times n + 1.008 \times 6 = 30.068$

$12.01 \times n = 30.068 - 1.008 \times 6$

$n = \dfrac{30.068 - 1.008 \times 6}{12.01}$

$n = \dfrac{30.068 - 6.048}{12.01}$

$n = \dfrac{24.02}{12.01}$

$n = 2$

There are 2 C atoms in a molecule of C_nH_6.

b.

n number of C atoms in a molecule of C_nH_6 (1)

$30.068 = 12.01n + 1.008 \times 6$

$12.01n + 1.008 \times 6 = 30.068$

$12.01n + 6.048 = 30.068$

$12.01n = 30.068 - 6.048$

$12.01n = 24.02$

$n = \dfrac{24.02}{12.01}$

$n = 2$

There are 2 C atoms in a molecule of C_nH_6.

16. a.

M atomic mass of the unknown atom, X (amu/atom)

$2690.94 = (32.07 + M \times 2) \times 42$

$(32.07 + M \times 2) \times 42 = 2690.94$

$32.07 + M \times 2 = \dfrac{2690.94}{42}$

$M \times 2 = \dfrac{2690.94}{42} - 32.07$

$M = \dfrac{\frac{2690.94}{42} - 32.07}{2}$

$M = \dfrac{64.07 - 32.07}{2}$

$M = \dfrac{32}{2}$

$M = 16$ amu/atom

The atomic mass of X is 16 amu/atom which identifies the atom as oxygen.

b.

M atomic mass of the unknown atom, X (amu/atom)

$2690.94 = 42\,(32.07 + 2M)$

$42\,(32.07 + 2M) = 2690.94$

$32.07 + 2M = \dfrac{2690.94}{42}$

$32.07 + 2M = 64.07$

$2M = 64.07 - 32.07$

$2M = 32$

$M = \dfrac{32}{2}$

$M = 16$ amu/atom

The atomic mass of X is 16 amu/atom which identifies the atom as oxygen.

17. a.

M molar mass of the unknown atom, X (g/mol)

$2690.94 = (M + 16.00 \times 2) \times 42$

$(M + 16.00 \times 2) \times 42 = 2690.94$

$M + 16.00 \times 2 = \dfrac{2690.94}{42}$

$M = \dfrac{2690.94}{42} - 16.00 \times 2$

$M = 64.07 - 32$

$M = 32.07$

The molar mass of X is 32.07 g/mol which identifies the atom as sulfur.

b.

M molar mass of the unknown atom, X (g/mol)

$2690.94 = 42\,(M + 16.00 \times 2)$

$42\,(M + 16.00 \times 2) = 2690.94$

$42\,(M + 32) = 2690.94$

$M + 32 = \dfrac{2690.94}{42}$

$M + 32 = 64.07$

$M = 64.07 - 32$

$M = 32.07$

The molar mass of X is 32.07 g/mol which identifies the atom as sulfur.

18. a.

M atomic mass of the unknown atom, X (amu/atom)

$74.096 = 40.08 + (M + 1.008) \times 2$

$40.08 + (M + 1.008) \times 2 = 74.096$

$(M + 1.008) \times 2 = 74.096 - 40.08$

$M + 1.008 = \dfrac{74.096 - 40.08}{2}$

$M = \dfrac{74.096 - 40.08}{2} - 1.008$

$M = \dfrac{34.016}{2} - 1.008$

$M = 17.008 - 1.008$

$M = 16$

The atomic mass of X is 16 amu/atom which identifies the atom as oxygen.

b.

M atomic mass of the unknown atom, X (amu/atom)

$74.096 = 40.08 + 2(M + 1.008)$

$40.08 + 2(M + 1.008) = 74.096$

$2(M + 1.008) = 74.096 - 40.08$

$2(M + 1.008) = 34.016$

$M + 1.008 = \dfrac{34.016}{2}$

$M + 1.008 = 17.008$

$M = 17.008 - 1.008$

$M = 16$

The atomic mass of X is 16 amu/atom which identifies the atom as oxygen.

19. a.

M molar mass of the unknown atom, X (g/mol)

$253.922 = M \times 2 + (12.01 + 16.00 \times 3) \times 4$

$M \times 2 + (12.01 + 16.00 \times 3) \times 4 = 253.922$

$M \times 2 = 253.922 - (12.01 + 16.00 \times 3) \times 4$

$M = \dfrac{253.922 - (12.01 + 16.00 \times 3) \times 4}{2}$

$M = \dfrac{253.922 - (12.01 + 48) \times 4}{2}$

$M = \dfrac{253.922 - 60.01 \times 4}{2}$

$M = \dfrac{253.922 - 240.04}{2}$

$M = \dfrac{13.882}{2}$

$M = 6.941 \text{ g/mol}$

The molar mass of X is 6.941 g/mol which identifies the atom as lithium.

b.

M molar mass of the unknown atom, X (g/mol)

$253.922 = 2M + 4(12.01 + 16.00 \times 3)$

$2M + 4(12.01 + 16.00 \times 3) = 253.922$

$2M + 4(12.01 + 48) = 253.922$

$2M + 4 \times 60.01 = 253.922$

$2M + 240.04 = 253.922$

$2M = 253.922 - 240.04$

$2M = 13.882$

$M = \dfrac{13.882}{2}$

$M = 6.941$ g/mol

The molar mass of X is 6.941 g/mol which identifies the atom as lithium.

20. a.

M molar mass of the unknown atom, X (g/mol)

$1808.976 = \left[22.99 \times 4 + (M + 16.00 \times 4) \times 3 \right] \times 4.8$

$\left[22.99 \times 4 + (M + 16.00 \times 4) \times 3 \right] \times 4.8 = 1808.976$

$22.99 \times 4 + (M + 16.00 \times 4) \times 3 = \dfrac{1808.976}{4.8}$

$(M + 16.00 \times 4) \times 3 = \dfrac{1808.976}{4.8} - 22.99 \times 4$

$M + 16.00 \times 4 = \dfrac{\dfrac{1808.976}{4.8} - 22.99 \times 4}{3}$

$M = \dfrac{\dfrac{1808.976}{4.8} - 22.99 \times 4}{3} - 16.00 \times 4$

$M = \dfrac{376.87 - 91.96}{3} - 64$

$M = \dfrac{284.91}{3} - 64$

$M = 94.97 - 64$

$M = 30.97$ g/mol

The molar mass of X is 30.97 g/mol which identifies the atom as phosphorus.

b.

M molar mass of the unknown atom, X (g/mol)

$$1808.976 = 4.8\left[22.99 \times 4 + 3\left(M + 16.00 \times 4\right)\right]$$

$$4.8\left[22.99 \times 4 + 3\left(M + 16.00 \times 4\right)\right] = 1808.976$$

$$4.8\left[91.96 + 3\left(M + 64\right)\right] = 1808.976$$

$$91.96 + 3\left(M + 64\right) = \frac{1808.976}{4.8}$$

$$91.96 + 3\left(M + 64\right) = 376.87$$

$$3\left(M + 64\right) = 376.87 - 91.96$$

$$3\left(M + 64\right) = 284.91$$

$$M + 64 = \frac{284.91}{3}$$

$$M + 64 = 94.97$$

$$M = 94.97 - 64$$

$$M = 30.97 \text{ g/mol}$$

The molar mass of X is 30.97 g/mol which identifies
the atom as phosphorus.

21. a.

n number of molecules of $(NH_4)_3PO_4$ (1)

$$15\,058.696 = \left[\left(14.01 + 1.008 \times 4\right) \times 3 + 30.97 + 16.00 \times 4\right] \times n$$

$$\left[\left(14.01 + 1.008 \times 4\right) \times 3 + 30.97 + 16.00 \times 4\right] \times n = 15\,058.696$$

$$n = \frac{15\,058.696}{\left(14.01 + 1.008 \times 4\right) \times 3 + 30.97 + 16.00 \times 4}$$

$$n = \frac{15\,058.696}{\left(14.01 + 4.032\right) \times 3 + 30.97 + 64}$$

$$n = \frac{15\,058.696}{18.042 \times 3 + 30.97 + 64}$$

$$n = \frac{15\,058.696}{54.126 + 30.97 + 64}$$

$$n = \frac{15\,058.696}{149.096}$$

$$n = 101$$

There are 101 molecules of $(NH_4)_3PO_4$ in the sample.

b.

n number of molecules of $(NH_4)_3PO_4$ (1)

$15\,058.696 = \left[3\,(14.01 + 1.008 \times 4) + 30.97 + 16.00 \times 4\right]n$

$\left[3\,(14.01 + 1.008 \times 4) + 30.97 + 16.00 \times 4\right]n = 15\,058.696$

$\left[3\,(14.01 + 4.032) + 30.97 + 64\right]n = 15\,058.696$

$(3 \times 18.042 + 30.97 + 64)\,n = 15\,058.696$

$(54.126 + 30.97 + 64)\,n = 15\,058.696$

$149.096n = 15\,058.696$

$n = \dfrac{15\,058.696}{149.096}$

$n = 101$

There are 101 molecules of $(NH_4)_3PO_4$ in the sample.

22. a.

t temperature on the third day (°C)

$15.8 = \dfrac{12 + 15 + t + 14 + 20}{5}$

$\dfrac{12 + 15 + t + 14 + 20}{5} = 15.8$

$12 + 15 + t + 14 + 20 = 5 \times 15.8$

$t = 5 \times 15.8 - 12 - 15 - 14 - 20$

$t = 79 - 12 - 15 - 14 - 20$

$t = 18\ °C$

The temperature was 18 °C on the third day.

b.

t temperature on the third day (°C)

$15.8 = \dfrac{12 + 15 + t + 14 + 20}{5}$

$\dfrac{12 + 15 + t + 14 + 20}{5} = 15.8$

$\dfrac{61 + t}{5} = 15.8$

$61 + t = 5 \times 15.8$

$61 + t = 79$

$t = 79 - 61$

$t = 18\ °C$

The temperature was 18 °C on the third day.

23. a.

a maximum amount of money Marie can spend ($)

$$2000 = \frac{2100 + 1600 + a}{3}$$

$$\frac{2100 + 1600 + a}{3} = 2000$$

$$2100 + 1600 + a = 3 \times 2000$$

$$a = 3 \times 2000 - 2100 - 1600$$

$$a = 6000 - 2100 - 1600$$

$$a = \$2300$$

Marie can spend at most $2300 this month.

b.

a maximum amount of money Marie can spend ($)

$$2000 = \frac{2100 + 1600 + a}{3}$$

$$\frac{2100 + 1600 + a}{3} = 2000$$

$$\frac{3700 + a}{3} = 2000$$

$$3700 + a = 3 \times 2000$$

$$3700 + a = 6000$$

$$a = 6000 - 3700$$

$$a = \$2300$$

Marie can spend at most $2300 this month.

24. a.

c cost of a Type C computer system ($/system)

$$1043.33 = \frac{1200 \times 8 + 820 + c \times 3}{8 + 1 + 3}$$

$$\frac{1200 \times 8 + 820 + c \times 3}{8 + 1 + 3} = 1043.33$$

$$1200 \times 8 + 820 + c \times 3 = (8 + 1 + 3) \times 1043.33$$

$$c \times 3 = (8 + 1 + 3) \times 1043.33 - 1200 \times 8 - 820$$

$$c = \frac{(8 + 1 + 3) \times 1043.33 - 1200 \times 8 - 820}{3}$$

$$c = \frac{12 \times 1043.33 - 9600 - 820}{3}$$

$$c = \frac{12\,519.96 - 9600 - 820}{3}$$

$$c = \frac{2099.96}{3}$$

$$c = \$699.99$$

The cost of a Type C computer system is $699.99.

b.

c cost of a Type C computer system (\$/system)

$$1043.33 = \frac{1200 \times 8 + 820 + 3c}{8 + 1 + 3}$$

$$\frac{1200 \times 8 + 820 + 3c}{8 + 1 + 3} = 1043.33$$

$$\frac{9600 + 820 + 3c}{12} = 1043.33$$

$$\frac{10\,420 + 3c}{12} = 1043.33$$

$$10\,420 + 3c = 12 \times 1043.33$$

$$10\,420 + 3c = 12\,519.96$$

$$3c = 12\,519.96 - 10\,420$$

$$3c = 2099.96$$

$$c = \frac{2099.96}{3}$$

$$c = \$699.99$$

The cost of a Type C computer system is \$699.99.

25. a.

r rate of pay at my third job (\$/h)

$$15.04 = \frac{14 \times 15 + 16 \times 10 + r \times 2}{15 + 10 + 2}$$

$$\frac{14 \times 15 + 16 \times 10 + r \times 2}{15 + 10 + 2} = 15.04$$

$$14 \times 15 + 16 \times 10 + r \times 2 = (15 + 10 + 2) \times 15.04$$

$$r \times 2 = (15 + 10 + 2) \times 15.04 - 14 \times 15 - 16 \times 10$$

$$r = \frac{(15 + 10 + 2) \times 15.04 - 14 \times 15 - 16 \times 10}{2}$$

$$r = \frac{27 \times 15.04 - 210 - 160}{2}$$

$$r = \frac{406.08 - 210 - 160}{2}$$

$$r = \frac{36.08}{2}$$

$$r = \$18.04/h$$

My rate of pay at my third job is \$18.04/h.

b.

r rate of pay at my third job ($/h)

$$15.04 = \frac{14 \times 15 + 16 \times 10 + 2r}{15 + 10 + 2}$$

$$\frac{14 \times 15 + 16 \times 10 + 2r}{15 + 10 + 2} = 15.04$$

$$\frac{210 + 160 + 2r}{27} = 15.04$$

$$\frac{370 + 2r}{27} = 15.04$$

$$370 + 2r = 27 \times 15.04$$

$$370 + 2r = 406.08$$

$$2r = 406.08 - 370$$

$$2r = 36.08$$

$$r = \frac{36.08}{2}$$

$$r = \$18.04/h$$

My rate of pay at my third job is $18.04/h.

26. a.

r training time per week per worker in Section C (h/week/worker)

$$4 = \frac{3 \times 120 + 5 \times 250 + r \times 320}{120 + 250 + 320}$$

$$\frac{3 \times 120 + 5 \times 250 + r \times 320}{120 + 250 + 320} = 4$$

$$3 \times 120 + 5 \times 250 + r \times 320 = (120 + 250 + 320) \times 4$$

$$r \times 320 = (120 + 250 + 320) \times 4 - 3 \times 120 - 5 \times 250$$

$$r = \frac{(120 + 250 + 320) \times 4 - 3 \times 120 - 5 \times 250}{320}$$

$$r = \frac{690 \times 4 - 360 - 1250}{320}$$

$$r = \frac{2760 - 360 - 1250}{320}$$

$$r = \frac{1150}{320}$$

$$r = 3.593\,75$$

Each worker in Section C should receive 3.6 h of training per week.

b.

r training time per week per worker in Section C (h/week/worker)

$$4 = \frac{3 \times 120 + 5 \times 250 + 320r}{120 + 250 + 320}$$

$$\frac{3 \times 120 + 5 \times 250 + 320r}{120 + 250 + 320} = 4$$

$$\frac{360 + 1250 + 320r}{690} = 4$$

$$\frac{1610 + 320r}{690} = 4$$

$$1610 + 320r = 690 \times 4$$

$$1610 + 320r = 2760$$

$$320r = 2760 - 1610$$

$$320r = 1150$$

$$r = \frac{1150}{320}$$

$$r = 3.593\,75$$

Each worker in Section C should receive 3.6 h of training per week.

27. a.

g grade point in CHEM101 for a GPA of 3.00 (1)

$$3.00 = \frac{3.33 \times 3 + 1.67 \times 5 + g \times 4}{3 + 5 + 4}$$

$$\frac{3.33 \times 3 + 1.67 \times 5 + g \times 4}{3 + 5 + 4} = 3.00$$

$$3.33 \times 3 + 1.67 \times 5 + g \times 4 = (3 + 5 + 4) \times 3.00$$

$$g \times 4 = (3 + 5 + 4) \times 3.00 - 3.33 \times 3 - 1.67 \times 5$$

$$g = \frac{(3 + 5 + 4) \times 3.00 - 3.33 \times 3 - 1.67 \times 5}{4}$$

$$g = \frac{12 \times 3.00 - 9.99 - 8.35}{4}$$

$$g = \frac{36 - 9.99 - 8.35}{4}$$

$$g = \frac{17.66}{4}$$

$$g = 4.415$$

I will need a grade point value of 4.42 in CHEM101
for a GPA of 3.00 which is not possible.

b.

g grade point in CHEM101 for a GPA of 3.00 (1)

$$3.00 = \frac{3.33 \times 3 + 1.67 \times 5 + 4g}{3 + 5 + 4}$$

$$\frac{3.33 \times 3 + 1.67 \times 5 + 4g}{3 + 5 + 4} = 3.00$$

$$\frac{9.99 + 8.35 + 4g}{12} = 3.00$$

$$\frac{18.34 + 4g}{12} = 3.00$$

$$18.34 + 4g = 12 \times 3.00$$

$$18.34 + 4g = 36$$

$$4g = 36 - 18.34$$

$$4g = 17.66$$

$$g = \frac{17.66}{4}$$

$$g = 4.415$$

I will need a grade point value of 4.42 in CHEM101 for a GPA of 3.00 which is not possible.

28. a.

g grade point in COMM1020 for a GPA of 3.40 (1)

$$3.40 = \frac{4.00 \times 3 + 2.67 \times 2 + 2.33 \times 4 + g \times 3}{3 + 2 + 4 + 3}$$

$$\frac{4.00 \times 3 + 2.67 \times 2 + 2.33 \times 4 + g \times 3}{3 + 2 + 4 + 3} = 3.40$$

$$4.00 \times 3 + 2.67 \times 2 + 2.33 \times 4 + g \times 3 = (3 + 2 + 4 + 3) \times 3.40$$

$$g \times 3 = (3 + 2 + 4 + 3) \times 3.40 - 4.00 \times 3 - 2.67 \times 2 - 2.33 \times 4$$

$$g = \frac{(3 + 2 + 4 + 3) \times 3.40 - 4.00 \times 3 - 2.67 \times 2 - 2.33 \times 4}{3}$$

$$g = \frac{12 \times 3.40 - 12 - 5.34 - 9.32}{3}$$

$$g = \frac{40.8 - 12 - 5.34 - 9.32}{3}$$

$$g = \frac{14.14}{3}$$

$$g = 4.713 \ldots$$

I will need a grade point value of 4.71 in COMM1020 for a GPA of 3.40 which is impossible.

b.

g grade point in COMM1020 for a GPA of 3.40 (1)

$$3.40 = \frac{4.00 \times 3 + 2.67 \times 2 + 2.33 \times 4 + 3g}{3 + 2 + 4 + 3}$$

$$\frac{4.00 \times 3 + 2.67 \times 2 + 2.33 \times 4 + 3g}{3 + 2 + 4 + 3} = 3.40$$

$$\frac{12 + 5.34 + 9.32 + 3g}{12} = 3.40$$

$$\frac{26.66 + 3g}{12} = 3.40$$

$$26.66 + 3g = 12 \times 3.40$$

$$26.66 + 3g = 40.8$$

$$3g = 40.8 - 26.66$$

$$3g = 14.14$$

$$g = \frac{14.14}{3}$$

$$g = 4.713\, ldots$$

I will need a grade point value of 4.71 in COMM1020 for a GPA of 3.40 which is impossible.

29. a.

p price of the hat ($)

$$4.50 = 0.2 \times p$$

$$0.2 \times p = 4.50$$

$$p = \frac{4.50}{0.2}$$

$$p = \$22.5$$

The hat is priced at $22.50.

b.

p price of the hat ($)

$$4.50 = 0.2p$$

$$0.2p = 4.50$$

$$p = \frac{4.50}{0.2}$$

$$p = \$22.5$$

The hat is priced at $22.50.

30. a.

r rate of discount (1)

$16.38 = (1 - r) \times 18.20$

$(1 - r) \times 18.20 = 16.38$

$1 - r = \dfrac{16.38}{18.20}$

$-r = -1 + \dfrac{16.38}{18.20}$

$r = 1 - \dfrac{16.38}{18.20}$

$r = 1 - 0.9$

$r = 0.1$

The rate of discount is 10%.

b.

r rate of discount (1)

$16.38 = 18.20\,(1 - r)$

$18.20\,(1 - r) = 16.38$

$1 - r = \dfrac{16.38}{18.20}$

$1 - r = 0.9$

$-r = 0.9 - 1$

$-r = -0.1$

$r = 0.1$

The rate of discount is 10%.

31. a.

p price of the history textbook ($)

$3.38 = 0.13 \times p$

$0.13 \times p = 3.38$

$p = \dfrac{3.38}{0.13}$

$p = 26$

The history textbook is priced at $26.00.

b.

p price of the history textbook ($)

$3.38 = 0.13p$

$0.13p = 3.38$

$p = \dfrac{3.38}{0.13}$

$p = 26$

The history textbook is priced at $26.00.

32. a.

p price of the gloves ($)

$17.48 = (1 + 0.13) \times (1 - 0.15) \times p$

$(1 + 0.13) \times (1 - 0.15) \times p = 17.48$

$p = \dfrac{17.48}{(1 + 0.13) \times (1 - 0.15)}$

$p = \dfrac{17.48}{1.13 \times 0.85}$

$p = 18.198 \ldots$

$p = \$18.20$

The gloves are priced at $18.20.

b.

p price of the gloves ($)

$17.48 = (1 + 0.13)(1 - 0.15)p$

$(1 + 0.13)(1 - 0.15)p = 17.48$

$1.13(0.85)p = 17.48$

$0.9605p = 17.48$

$p = \dfrac{17.48}{0.9605}$

$p = 18.198 \ldots$

$p = \$18.20$

The gloves are priced at $18.20.

33. a.

r rate of tax (1)

$17.48 = (1 + r) \times (1 - 0.15) \times 18.20$

$(1 + r) \times (1 - 0.15) \times 18.20 = 17.48$

$1 + r = \dfrac{17.48}{(1 - 0.15) \times 18.20}$

$r = \dfrac{17.48}{(1 - 0.15) \times 18.20} - 1$

$r = \dfrac{17.48}{0.85 \times 18.20} - 1$

$r = 1.129 \ldots - 1$

$r = 0.129 \ldots$

$r = 0.13$

The rate of tax is 13%.

b.

r rate of tax (1)

$17.48 = (1 - 0.15)(18.20)(1 + r)$

$(1 - 0.15)(18.20)(1 + r) = 17.48$

$0.85(18.20)(1 + r) = 17.48$

$15.47(1 + r) = 17.48$

$1 + r = \dfrac{17.48}{15.47}$

$1 + r = 1.129\ldots$

$r = 1.129\ldots - 1$

$r = 0.129\ldots$

$r = 0.13$

The rate of tax is 13%.

34. a.

p price of the laptop ($)

$132.60 = 0.13 \times (1 - 0.15) \times p$

$0.13 \times (1 - 0.15) \times p = 132.60$

$p = \dfrac{132.60}{0.13 \times (1 - 0.15)}$

$p = \dfrac{132.60}{0.13 \times 0.85}$

$p = \$1200$

The laptop is priced at $1200.00.

b.

p price of the laptop ($)

$132.60 = 0.13(1 - 0.15)p$

$0.13(1 - 0.15)p = 132.60$

$0.13(0.85)p = 132.60$

$0.1105p = 132.60$

$p = \dfrac{132.60}{0.1105}$

$p = \$1200$

The laptop is priced at $1200.00.

35. a.

r ⠀⠀rate of discount (1)

$2.90 = 0.13 \times (1 - r) \times 24.80$

$0.13 \times (1 - r) \times 24.80 = 2.90$

$1 - r = \dfrac{2.90}{0.13 \times 24.80}$

$-r = -1 + \dfrac{2.90}{0.13 \times 24.80}$

$r = 1 - \dfrac{2.90}{0.13 \times 24.80}$

$r = 1 - 0.899 \ldots$

$r = 0.100 \ldots$

$r = 0.1$

The rate of discount is 10%.

b.

r ⠀⠀rate of discount (1)

$2.90 = 0.13\,(24.80)\,(1 - r)$

$0.13\,(24.80)\,(1 - r) = 2.90$

$3.224\,(1 - r) = 2.90$

$1 - r = \dfrac{2.90}{3.224}$

$1 - r = 0.899 \ldots$

$-r = 0.899 \ldots - 1$

$-r = -0.100 \ldots$

$r = 0.100 \ldots$

$r = 0.1$

The rate of discount is 10%.

36. a.

r ⠀⠀rate of tip (1)

$3.00 = r \times 19.10$

$r \times 19.10 = 3.00$

$r = \dfrac{3.00}{19.10}$

$r = 0.157 \ldots$

$r = 0.16$

The rate of tip was 16%.

b.

r rate of tip (1)

$3.00 = 19.10r$

$19.10r = 3.00$

$r = \dfrac{3.00}{19.10}$

$r = 0.157\ldots$

$r = 0.16$

The rate of tip was 16%.

37. a.

n number of respondents (1)

$1246 = 0.82 \times n$

$0.82 \times n = 1246$

$n = \dfrac{1246}{0.82}$

$n = 1519.512\ldots$

$n = 1519$ respondents

There were 1519 respondents.

b.

n number of respondents (1)

$1246 = 0.82n$

$0.82n = 1246$

$n = \dfrac{1246}{0.82}$

$n = 1519.512\ldots$

$n = 1519$ respondents

There were 1519 respondents.

38. a.

n number of respondents (1)

$274 = (1 - 0.82) \times n$

$(1 - 0.82) \times n = 274$

$n = \dfrac{274}{1 - 0.82}$

$n = \dfrac{274}{0.18}$

$n = 1522.222\ldots$

$n = 1522$

There were 1522 respondents.

b.

> n number of respondents (1)
>
> $274 = (1 - 0.82)\,n$
>
> $(1 - 0.82)\,n = 274$
>
> $0.18n = 274$
>
> $n = \dfrac{274}{0.18}$
>
> $n = 1522.222\ldots$
>
> $n = 1522$
>
> There were 1522 respondents.

39. a.

> r percent rise in population (1)
>
> $5544 = r \times 132\,000$
>
> $r \times 132\,000 = 5544$
>
> $r = \dfrac{5544}{132\,000}$
>
> $r = 0.042$
>
> The population rose by 4.2 %.

b.

> r percent rise in population (1)
>
> $5544 = 132\,000r$
>
> $132\,000r = 5544$
>
> $r = \dfrac{5544}{132\,000}$
>
> $r = 0.042$
>
> The population rose by 4.2 %.

40. a.

> n population of the town ten years ago (1)
>
> $5544 = 0.042 \times n$
>
> $0.042 \times n = 5544$
>
> $n = \dfrac{5544}{0.042}$
>
> $n = 132\,000$ people
>
> The population of the town was 132 000 ten years ago.

b.

n population of the town ten years ago (1)

$5544 = 0.042n$

$0.042n = 5544$

$n = \dfrac{5544}{0.042}$

$n = 132\,000$ people

The population of the town was $132\,000$ ten years ago.

41. a.

n number of eligible voters (1)

$76 = 0.18 \times 0.78 \times n$

$0.18 \times 0.78 \times n = 76$

$n = \dfrac{76}{0.18 \times 0.78}$

$n = 541.310 \ldots$

$n = 541$ eligible voters

There are 541 eligible voters.

b.

n number of eligible voters (1)

$76 = 0.18\,(0.78)\,n$

$0.18\,(0.78)\,n = 76$

$0.1404n = 76$

$n = \dfrac{76}{0.1404}$

$n = 541.310 \ldots$

$n = 541$ eligible voters

There are 541 eligible voters.

42. a.

n number of eligible voters (1)

$346 = (1 - 0.18) \times 0.78 \times n$

$(1 - 0.18) \times 0.78 \times n = 346$

$n = \dfrac{346}{(1 - 0.18) \times 0.78}$

$n = \dfrac{346}{0.82 \times 0.78}$

$n = 540.963 \ldots$

$n = 541$ eligible voters

There are 541 eligible voters.

b.

n number of eligible voters (1)

$346 = (1 - 0.18)(0.78)n$

$(1 - 0.18)(0.78)n = 346$

$0.82(0.78)n = 346$

$0.6396n = 346$

$n = \dfrac{346}{0.6396}$

$n = 540.963 \ldots$

$n = 541$ eligible voters

There are 541 eligible voters.

43. a.

r percentage of eligible voters who voted (1)

$76 = 0.18 \times r \times 541$

$0.18 \times r \times 541 = 76$

$r = \dfrac{76}{0.18 \times 541}$

$r = 0.780 \ldots$

78% of the eligible voters voted.

b.

r percentage of eligible voters who voted (1)

$76 = 0.18(541)r$

$0.18(541)r = 76$

$97.38r = 76$

$r = \dfrac{76}{97.38}$

$r = 0.780 \ldots$

78% of the eligible voters voted.

44. a.

r percentage of eligible voters who voted (1)

$346 = (1 - 0.18) \times r \times 541$

$(1 - 0.18) \times r \times 541 = 346$

$r = \dfrac{346}{(1 - 0.18) \times 541}$

$r = \dfrac{346}{0.82 \times 541}$

$r = 0.779 \ldots$

78% of the eligible voters voted.

b.

r percentage of eligible voters who voted (1)

$346 = (1 - 0.18)(541)r$

$(1 - 0.18)(541)r = 346$

$0.82(541)r = 346$

$443.62r = 346$

$r = \dfrac{346}{443.62}$

$r = 0.779 \ldots$

78% of the eligible voters voted.

45. a.

r percentage of eligible voters who voted
and opposed the proposed changes (1)

$76 = r \times 0.78 \times 541$

$r \times 0.78 \times 541 = 76$

$r = \dfrac{76}{0.78 \times 541}$

$r = 0.180 \ldots$

18% of eligible voters who voted and opposed the proposed changes.

b.

r percentage of eligible voters who voted
and opposed the proposed changes (1)

$76 = 0.78(541)r$

$0.78(541)r = 76$

$421.98r = 76$

$r = \dfrac{76}{421.98}$

$r = 0.180 \ldots$

18% of eligible voters who voted and opposed the proposed changes.

46. a.

r percentage of eligible voters who voted and opposed the proposed changes (1)

$346 = (1 - r) \times 0.78 \times 541$

$(1 - r) \times 0.78 \times 541 = 346$

$1 - r = \dfrac{346}{0.78 \times 541}$

$-r = -1 + \dfrac{346}{0.78 \times 541}$

$r = 1 - \dfrac{346}{0.78 \times 541}$

$r = 1 - 0.819 \ldots$

$r = 0.180 \ldots$

18% of eligible voters voted and opposed the proposed changes.

b.

r percentage of eligible voters who voted and opposed the proposed changes (1)

$346 = 0.78 \, (541) \, (1 - r)$

$0.78 \, (541) \, (1 - r) = 346$

$421.98 \, (1 - r) = 346$

$1 - r = \dfrac{346}{421.98}$

$1 - r = 0.819 \ldots$

$-r = -1 + 0.819 \ldots$

$-r = -0.180 \ldots$

$r = 0.180 \ldots$

18% of eligible voters voted and opposed the proposed changes.

47. a.

n number of frogs in the sample (1)

$62 = 0.15 \times 0.34 \times n$

$0.15 \times 0.34 \times n = 62$

$n = \dfrac{62}{0.15 \times 0.34}$

$n = 1215.686 \ldots$

$n = 1216$ frogs

There are 1216 frogs in the sample.

b.

n number of frogs in the sample (1)

$62 = 0.15\,(0.34)\,n$

$0.15\,(0.34)\,n = 62$

$0.051n = 62$

$n = \dfrac{62}{0.051}$

$n = 1215.686\,\ldots$

$n = 1216$ frogs

There are 1216 frogs in the sample.

48. a.

r percentage of participants who experienced side effects and needed to be hospitalized (1)

$70 = (1 - r) \times 0.61 \times 120$

$(1 - r) \times 0.61 \times 120 = 70$

$1 - r = \dfrac{70}{0.61 \times 120}$

$-r = -1 + \dfrac{70}{0.61 \times 120}$

$r = 1 - \dfrac{70}{0.61 \times 120}$

$r = 1 - 0.956\,\ldots$

$r = 0.043\,715\,\ldots$

4.4% of participants experienced side effects and needed to be hospitalized.

b.

r percentage of participants who experienced side effects and needed to be hospitalized (1)

$70 = 0.61\,(120)\,(1 - r)$

$0.61\,(120)\,(1 - r) = 70$

$73.2\,(1 - r) = 70$

$1 - r = \dfrac{70}{73.2}$

$1 - r = 0.956\,\ldots$

$-r = -1 + 0.956\,\ldots$

$-r = -0.043\,715\,\ldots$

$r = 0.043\,715\,\ldots$

4.4% of participants experienced side effects and needed to be hospitalized.

49. a.

r rate of decrease in population of Blue Lake City in the nineties (1)

$181\,440 = (1 - r) \times (1 + 0.12) \times 180\,000$

$(1 - r) \times (1 + 0.12) \times 180\,000 = 181\,440$

$1 - r = \dfrac{181\,440}{(1 + 0.12) \times 180\,000}$

$-r = -1 + \dfrac{181\,440}{(1 + 0.12) \times 180\,000}$

$r = 1 - \dfrac{181\,440}{(1 + 0.12) \times 180\,000}$

$r = 1 - \dfrac{181\,440}{1.12 \times 180\,000}$

$r = 1 - 0.9$

$r = 0.1$

The population Blue Lake City decreased by 10% in the nineties.

b.

r rate of decrease in population of Blue Lake City in the nineties (1)

$181\,440 = (1 + 0.12)\,(180\,000)\,(1 - r)$

$(1 + 0.12)\,(180\,000)\,(1 - r) = 181\,440$

$1.12\,(180\,000)\,(1 - r) = 181\,440$

$201\,600\,(1 - r) = 181\,440$

$1 - r = \dfrac{181\,440}{201\,600}$

$1 - r = 0.9$

$-r = 0.9 - 1$

$-r = -0.1$

$r = 0.1$

The population Blue Lake City decreased by 10% in the nineties.

50. a.

n number of people who commute to work daily (1)

$45\,000 = (1 - 0.72) \times (1 - 0.33) \times n$

$(1 - 0.72) \times (1 - 0.33) \times n = 45\,000$

$n = \dfrac{45\,000}{(1 - 0.72) \times (1 - 0.33)}$

$n = \dfrac{45\,000}{0.28 \times 0.67}$

$n = 239\,872.068 \ldots$

$n = 239\,872$ people

$239\,872$ people commute to work daily.

b.

n number of people who commute to work daily (1)

$$45\,000 = (1 - 0.72)\,(1 - 0.33)\,n$$
$$(1 - 0.72)\,(1 - 0.33)\,n = 45\,000$$
$$0.28\,(0.67)\,n = 45\,000$$
$$0.1876n = 45\,000$$
$$n = \frac{45\,000}{0.1876}$$
$$n = 239\,872.068\ldots$$
$$n = 239\,872 \text{ people}$$

$239\,872$ people commute to work daily.

Exercise Set 9.3B

1. To extract the steps in the corresponding step-by-step solution, we begin with the expression on the right side of the equation at the end of the logical stage. This is the expression

$$3 \times 80\% - (75\% + 81\%)$$

NUMBER OF STEPS
We begin by noting that there are three operations present in the expression above. Since the steps need to be taken individually, there will be three steps in the solution.

ANALYSIS
The expression analyzes into two terms. The first is the term $3 \times 80\%$ and the second is the term $(75\% + 81\%)$. To evaluate the expression, we need to evaluate these terms and then subtract the latter from the former.

SYNTHESIS
We now extract the steps from the solution above. For each step, we identify the subexpression in the expression above that corresponds to that step and then think about the objective of the corresponding computation. This objective will be stated at the beginning of each step.

The first step corresponds to the evaluation of the term $3 \times 80\%$. A bit of reflection shows that this computes the total of all the grades as 80% is the average grade and there are 3 tests in total.[11] The first step follows:

Step 1. Calculate the sum of all the grades.

$$3 \times 80\% = 240\%$$

The sum of the grades is 240%.

---

[11] Note that this is the value of the quantity that is represented by the term, $3 \times 80\%$.

The expression in the model above now becomes

$$240\% \ - \ (75\% \ + \ 81\%)$$

The second step corresponds to the evaluation of the term $(75\% + 81\%)$. A bit of reflection shows that this computes the sum of the grades on the tests that I have already written.[12] The second step follows:

Step 2. Calculate the sum of the grades on tests that I have written.

$$75\% \ + \ 81\% \ = \ 156\%$$

The sum of the grades on tests that I have written is 156%.

The expression in the model above now becomes

$$240\% \ - \ 156\%$$

The last step corresponds to the subtraction $240\% - 156\%$. A bit of reflection shows that this computes the value of the grade needed on Test 3 for the desired average as it subtracts the sum of the grades on Tests 1 and 2 from the sum of the grades on all three tests. The third step follows:

Step 3. Calculate the grade needed on Test 3 for the desired average.

$$240\% \ - \ 156\% \ = \ 84\%$$

I need 84% on Test 3 to get an average of 80%.

The expression in the model above now becomes

$$84\%$$

The full step-by-step solution without the intervening comments in given below.

A STEP BY STEP SOLUTION

Step 1. Calculate the sum of all the grades.

$$3 \times 80\% \ = \ 240\%$$

The sum of the grades is 240%.

Step 2. Calculate the sum of the grades on tests that I have written.

$$75\% \ + \ 81\% \ = \ 156\%$$

The sum of the grades on tests that I have written is 156%.

Step 3. Calculate the grade needed on Test 3 for the desired average.

$$240\% \ - \ 156\% \ = \ 84\%$$

I need 84% on Test 3 to get an average of 80%.

---

[12]Note that this is the value of the quantity that is represented by the term, $(75\% + 81\%)$.

2. To extract the steps in the corresponding step-by-step solution, we begin with the expression on the right side of the equation at the end of the logical stage. This is the expression

$$\frac{1048}{1 - 0.165}$$

NUMBER OF STEPS
Since there are two operations present in the expression above and they need to be taken individually, there will be two steps in the solution.

ANALYSIS
The expression analyzes into a single term containing a single factor. This factor involves a division. To work out the division, we need to work out the divisor, $1 - 0.165$, and then divide the dividend, 1048, by the divisor.

SYNTHESIS
We now extract the steps from the solution above. For each step, we identify the subexpression in the expression above that corresponds to that step and then think about the objective of the corresponding computation. This objective will be stated at the beginning of each step.

The first step corresponds to the evaluation of the divisor $1 - 0.165$. A bit of reflection shows that this, which maps onto $100\% - 16.5\%$, computes the percentage that relates to the number of non-red-eyed frogs: If 16.5% of the frogs in the sample have red eyes, then $100\% - 16.5\%$, or $1 - 0.165$ of them do not have red eyes. The first step follows:

Step 1. Calculate the percentage of the frogs that do not have red eyes.

$$1 - 0.165 = 0.835$$

83.5% of the frogs do not have red eyes.

The expression in the model above now becomes

$$\frac{1048}{0.835}$$

The second step corresponds to the division $\frac{1048}{0.835}$. A bit of reflection shows that this computes the number of frogs in the sample.[13] The second step follows:

Step 2. Calculate the number of frogs in the sample.

$$\frac{1048}{0.835} = 1255.089 \ldots$$

There are 1255 frogs in the sample.

---

[13] We will discuss this in more detail below.

The expression in the model above now becomes

1255

Let us take a closer look at the logic that flows through the solution above. We begin with the number of frogs. Since this is unknown, we will use the quantity symbol, n, to represent its value with. We write

n

Since the problem states that 16.5% of the frogs in the sample have red eyes, we draw a branch leading from the total number of frogs, n, to the number of red-eyed frogs. We mark the branch with $\times 0.165$ to remind us that we can find the number of red-eyed frogs in the sample, by multiplying n by 0.165. We have

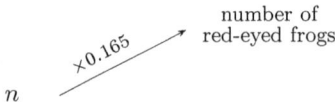

However, since the problem provides us with information on the number of non-red-eyed frogs (as opposed to the number of red-eyed frogs), we generate a new branch leading to the number of non-red-eyed frogs. The percentage that corresponds to this path is $1 - 0.165$: Since 16.5% of the forgs in the sample have red eyes, $100\% - 16.5\%$ of the frogs in the sample do not have red eyes. Note that this is what the divisor of $\frac{1048}{1-0.165}$ computes. We now have

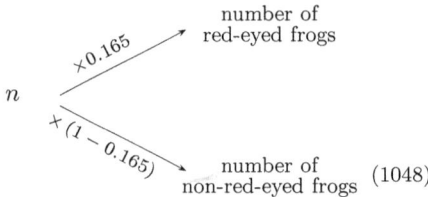

We now understand what the division of 1048 by $1 - 0.165$ computes: Since, moving forward along the lower branch in the diagram above requires that we multiply n by $1 - 0.165$, going back along the path requires that we divide 1048 by $1 - 0.165$. This is shown below.[14]

---

[14]Note that the forward path corresponds to the equation at the beginning of the logical stage, i.e.,

$$(1 - 0.165) \times n = 1048$$

$$n \xleftarrow{\div (1 - 0.165)} \underset{\substack{\text{number of} \\ \text{non-red-eyed frogs}}}{} \quad (1048)$$

The full step-by-step solution without the intervening comments in given below.

A STEP BY STEP SOLUTION

Step 1. Calculate the percentage that relates to the number of non-red-eyed frogs.

$$1 - 0.165 = 0.835$$

83.5% of the frogs in the sample do not have red eyes.

Step 2. Calculate the number of frogs in the sample.

$$1048 \div 0.835 = 1255.089 \ldots$$

There are 1255 frogs in the sample.

3. To extract the steps in the corresponding step-by-step solution, we begin with the expression on the right side of the equation at the end of the logical stage. This is the expression

$$\frac{320}{0.8 \times 0.9 \times 0.4 \times 1500}$$

NUMBER OF STEPS AND NUMBER OF SUB-STEPS PER STEP

The following diagram illustrates the problem posed in this exercise.

$$1500 \xrightarrow{\times 0.4} n_r \xrightarrow{\times 0.9} n_{r,g} \xrightarrow{\times r} n_{r,g,a} \xrightarrow{\times 0.8} 302$$

In this diagram, the value 1500 represents the given number of frogs in the sample and the value 302 represents the given number of red-eyed, green-skinned frogs in the sample that grow into adult frogs and have offspring. The quantity symbol n_r represents the number of red-eyed frogs in the sample, $n_{r,g}$ represents the number of red-eyed, green-skinned frogs in the sample, and $n_{r,g,a}$ represent the number of red-eyed, green-skinned frogs in the sample that grow into adult frogs.

Our objective is to find r, the rate that connects $n_{r,g}$ to $n_{r,g,a}$.

---

while the backward path corresponds to the equation at the end of the logical stage, i.e.,

$$n = \frac{1048}{1 - 0.165}$$

It is a good idea to keep in mind that the logic behind division by a percentage retraces a forward step in the manner that is illustrated by the diagrams above.

Since r is somewhere in the middle of the path (as opposed to the start or the end), the step-by-step solution will involve two major steps.

The first major step starts on the left with the value 1500, multiplies this by 0.4 to find n_r, and then multiplies n_r by 0.9 to calculate $n_{r,g}$. It stops when it reaches r. The first major step, then, will involve two sub-steps.

It is quite interesting that the first major step above corresponds to a move in the denominator of

$$\frac{302}{0.8 \times 0.9 \times 0.4 \times 1500}$$

starting on the right and moving back until we reach the place where r was located in the initial model, i.e., in between 0.8 and 0.9.

The second major step starts on the right with the value 302, divides this by 0.8 to find $n_{r,g,a}$. It stops when as it reaches r. The second major step, then, will involve a single sub-step.

Once we have sandwiched r from the left and the right, we divide the results to find r.

We provide the full step-by-step solution below and ask the reader to supply the rest of the explanations.

A STEP BY STEP SOLUTION

Step 1. Calculate the number of red-eyed, green skinned frogs.

     a. Calculate the number of red-eyed frogs.

$$0.4 \times 1500 \ = \ 600$$

     There are 600 red-eyed frogs in the sample.

     b. Calculate the number of red-eyed, green skinned frogs.

$$0.9 \times 600 \ = \ 540$$

     There are 540 red-eyed, green-skinned frogs in the sample.

Step 2. Calculate the number of red-eyed, green-skinned frogs that grow into adult frogs.

$$302 \div 0.8 \ = \ 377.5$$

     378 red-eyed, green-skinned frogs will grow into adult frogs.

Step 3. Calculate the percentage of red-eyed, green skinned frogs that grow into adult frogs.

$$r \ = \ 377.5 \div 540 \ = \ 0.699 \ldots$$

     70% of red-eyed, green-skinned frogs grow into adult frogs.

Note the manner in which proper ordering of the factors in the original model and the lines that follow it help us work out the corresponding step-by-step

125

solution from the given algebraic solution. Without this proper order,[15] one would still be able to extract a step-by-step solution from the algebraic solution but the associated logic would not be in line with the way we normally reason.

4. We will present the corresponding step-by-step solution below. We ask the reader to supply the detail behind the manner in which the relevant expression in the given algebraic solution maps onto the step-by-step solution below.

Step 1. Calculate the rate that applies to the grand total.

$$1 + 0.13 = 1.13$$

The grand total is equal to 113% of the sale price.

Step 2. Calculate the sale price.

$$14.60 \div 1.13 = 12.920\ldots$$

The sale price is $12.92.

Step 3. Calculate the rate that corresponds to the sale price.

$$12.920\ldots \div 15.20 = 0.850\ldots$$

The sale price is 85% of the price.

Step 4. Calculate the rate of discount.

$$1 - 0.850\ldots = 0.149\ldots$$

The rate of discount is 15%.

Exercise Set 10

1. a.
   b. Comparing the form of the equation $t_{\circ F} = \frac{9}{5}t_{\circ C} + 32$ with the general equation for linear equations, $y = mx + b$, sets m as $\frac{9}{5}$. This is the slope of the line.
   c. Comparing the form of the equation $t_{\circ F} = \frac{9}{5}t_{\circ C} + 32$ with the general equation for linear equations, $y = mx + b$, sets b as 32. The $t_{\circ F}$-intercept, then, is $(0, 32)$.
      For $t_{\circ C}$-intercept, we set $t_{\circ F} = 0$ and solve the equation for $t_{\circ C}$. This

---

[15] As one would see in such models as those generated through technical formulations.

$t_{{}^{\circ}F}$ (°F)

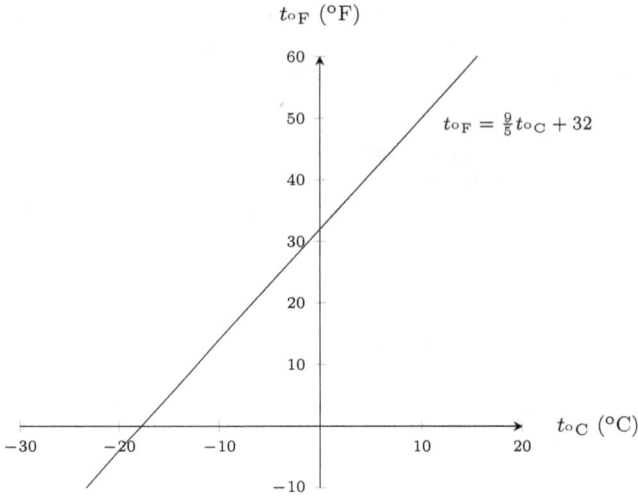

Figure 0.0.1: Graph of the equation $t_{{}^{\circ}F} = \frac{9}{5}t_{{}^{\circ}C} + 32$

yields

$$t_{{}^{\circ}F} = \frac{9}{5}t_{{}^{\circ}C} + 32$$

$$0 = \frac{9}{5}t_{{}^{\circ}C} + 32$$

$$\frac{9}{5}t_{{}^{\circ}C} + 32 = 0$$

$$\frac{9}{5}t_{{}^{\circ}C} = 0 - 32$$

$$t_{{}^{\circ}C} = \frac{5}{9}(0 - 32)$$

$$t_{{}^{\circ}C} = \frac{5}{9}(-32)$$

$$t_{{}^{\circ}C} = -17.777 \ldots {}^{\circ}C$$

The $t_{{}^{\circ}C}$-intercept is $(-17.8, 0)$

d. The slope of the line, i.e., $\frac{9\,{}^{\circ}F}{5\,{}^{\circ}C}$, relates the size of the units on the Celsius and Fahrenheit scales. It tells us that every 9 °F corresponds to 5 °C. We can turn this into the unit rate 1.8 °F/°C which states that every 1.8 °F corresponds to 1 °C.

The $t_{{}^{\circ}C}$-intercept represents the temperature at which brine (a solution made from ice, water and ammonium chloride) freezes in both °C and °F. The $t_{{}^{\circ}F}$-intercept represents the temperature at which water freezes in both °C and °F.

a ($)

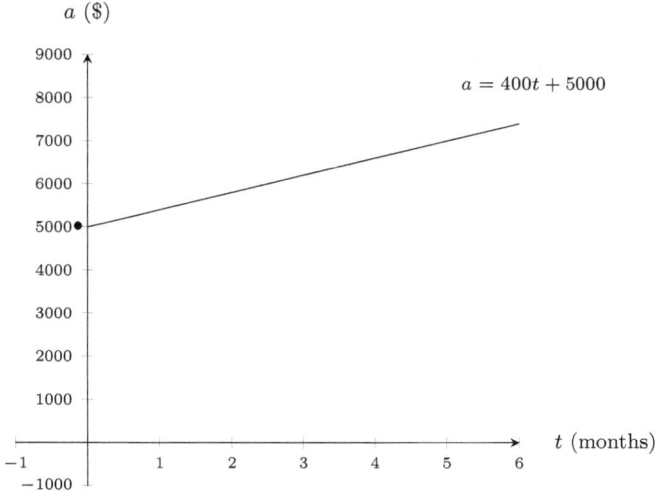

Figure 0.0.2: Graph of the equation $a = 400t + 5000$

2. a.

  b. Comparing the form of the equation $a = 400t + 5000$ with the general equation for linear equations, $y = mx + b$, sets m as 400. This is the slope of the line.

  c. Comparing the form of the equation $a = 400t + 5000$ with the general equation for linear equations, $y = mx + b$, sets b as 5000. The a-intercept, then, is $(0, 5000)$.

For t-intercept, we set $a = 0$ and solve the equation for t. This yields

$$a = 400t + 5000$$
$$0 = 400t + 5000$$
$$400t + 5000 = 0$$
$$400t = 0 - 5000$$
$$t = \frac{0 - 5000}{400}$$
$$t = \frac{-5000}{400}$$
$$t = -12.5$$

The t-intercept is $(-12.5, 0)$

  d. The slope of the line, i.e., \$400/month, relates the rate of monthly payments.

The a-intercept represents the amount of downpayment.

The function is defined only when t is greater than or equal to 0. Therefore, in the context of the present word problem, the t-intercept is undefined.

3. a. We will leave it to the reader to graph the equation (HINT: The graph is a straight line through the origin with a slope of 0.01 cm/m).

   b. Comparing the form of the equation $l_m = \frac{1}{100} l_{cm}$ with the general equation for linear equations, $y = mx + b$, sets m as $\frac{1}{100}$ or 0.01. This is the slope of the line.

   c. Since the graph is a straight line through the origin, both intercepts are $(0, 0)$.

   d. The slope of the line, i.e., $\frac{1\ m}{100\ cm}$ relates the size of the units metre and centimetre. It tells us that 1 m is equal to 100 cm.

   The common intercept of $(0, 0)$ indicates that the 0s on the two scales match, i.e., 0 m = 0 cm.

   e. The graph is a straight line through the origin.

Exercise Set 11.1

1. a. The concentration of OH^- ions decreases by a factor of $\frac{1}{10}$.
   b. The concentration of OH^- ions decreases by a factor of $\frac{1}{1000}$.
   c. The concentration of OH^- ions increases by a factor of 100.
   d. The concentration of H^+ ions decreases by a factor of $\frac{1}{1000}$.
   e. The concentration of H^+ ions decreases by a factor of $\frac{1}{4}$.

2. a. The concentration of H^+ ions in the solution decreases by a factor of $\frac{1}{10}$.
   b. The concentration of H^+ ions in the solution increases by a factor of 100.
   c. The concentration of OH^- ions in the solution decreases by a factor of $\frac{1}{1000}$.
   d. The concentration of OH^- ions in the solution increases by a factor of 10 000.

3. a. The pOH value of the solution decreases by 1.
   b. The pOH value of the solution increases by 3.
   c. The pH value of the solution decreases by 2.
   d. The pH value of the solution increases by 3.

4. a. The solution is basic as its pH value is greater than 7.
   b. The solution is acidic as its pH value is less than 7.
   c. The solution is neutral as its pH value is equal to 7.
   d. The solution is basic as its pOH value is less than 7.
   e. The solution is neutral as its pOH value is equal to 7.
   f. The solution is acidic as its pOH value is larger than 7.

5. a.

[H$^+$]   concentration of H$^+$ ions in the solution (mol/L)
[OH$^-$]   concentration of OH$^-$ ions in the solution (mol/L)
K_w dissociation constant for water (mol^2/L^2)

$$[H^+][OH^-] = K_w$$

$$[OH^-] = \frac{K_w}{[H^+]}$$

$$[OH^-] = \frac{1 \times 10^{-14}}{4.5 \times 10^{-4}}$$

$$[OH^-] = 2.222 \ldots \times 10^{-11} \text{ mol/L}$$

The concentration of OH$^-$ ions in the solution is 2.22×10^{-11} mol/L.

b.

[H$^+$]   concentration of H$^+$ ions in the solution (mol/L)
[OH$^-$]   concentration of OH$^-$ ions in the solution (mol/L)
K_w dissociation constant for water (mol^2/L^2)

$$[H^+][OH^-] = K_w$$

$$[H^+] = \frac{K_w}{[OH^-]}$$

$$[H^+] = \frac{1 \times 10^{-14}}{7.2 \times 10^{-8}}$$

$$[H^+] = 1.388 \ldots \times 10^{-7} \text{ mol/L}$$

The concentration of H$^+$ ions in the solution is 1.39×10^{-7} mol/L.

6. a.

[H$^+$]   concentration of H$^+$ ions in the solution (mol/L)
pH   the solution's power of hydrogen (1)

$$[H^+] = 10^{-\text{pH}}$$

$$[H^+] = 10^{-6.7}$$

$$[H^+] = 1.995 \ldots \times 10^{-7} \text{ mol/L}$$

The concentration of H$^+$ ions in the solution is 2.00×10^{-7} mol/L.

b.

[OH$^-$]   concentration of OH$^-$ ions in the solution (mol/L)
pOH   the solution's power of hydroxide (1)

$$[OH^-] = 10^{-\text{pOH}}$$

$$[OH^-] = 10^{-3.8}$$

$$[OH^-] = 1.584 \ldots \times 10^{-4} \text{ mol/L}$$

The concentration of OH$^-$ ions in the solution is 1.58×10^{-4} mol/L.

7. a.

$[H^+]$ concentration of H^+ ions in the solution (mol/L)
pH   the solution's power of hydrogen (1)

$[H^+] = 10^{-pH}$

$10^{-pH} = [H^+]$

$-pH = \log[H^+]$

$pH = -\log[H^+]$

$pH = -\log(6.1 \times 10^{-9})$

$pH = -(-8.214\ldots)$

$pH = 8.214\ldots$

The solution has a pH value of 8.2.

b.

$[OH^-]$ concentration of OH^- ions in the solution (mol/L)
pOH   the solution's power of hydroxide (1)

$[OH^-] = 10^{-pOH}$

$10^{-pOH} = [OH^-]$

$-pOH = \log[OH^-]$

$pOH = -\log[OH^-]$

$pOH = -\log(1.4 \times 10^{-12})$

$pOH = -(-11.853\ldots)$

$pOH = 11.853\ldots$

The solution has a pOH value of 11.9.

8. a.

pH   the solution's power of hydrogen (1)
pOH   the solution's power of hydroxide (1)
K_w dissociation constant for water (mol^2/L^2)

$pH + pOH = -\log K_w$

$pOH = -\log K_w - pH$

$pOH = -\log 10^{-14} - 3.4$

$pOH = -(-14) - 3.4$

$pOH = 14 - 3.4$

$pOH = 10.6$

The solution has a pOH value of 10.6.

b.

pH   the solution's power of hydrogen (1)
pOH   the solution's power of hydroxide (1)
K_w dissociation constant for water (mol^2/L^2)

$pH + pOH = -\log K_w$

$pH = -\log K_w - pOH$

$pH = -\log 10^{-14} - 9.7$

$pH = -(-14) - 9.7$

$pH = 14 - 9.7$

$pH = 4.3$

The solution has a pH value of 4.3.

Exercise Set 11.2

1.

m_0 initial mass of radioisotope (g)
m final mass of radioisotope (g)
n number of half-lives passed (1)

$m = \dfrac{1}{2^n} m_0$

$m = \dfrac{1}{2^{3.5}} \times 8.7$

$m = \dfrac{1}{11.313 \ldots} \times 8.7$

$m = 0.768 \ldots$ g

0.769 g of the radioisotope will be left.

2.

m_0 initial mass of radioisotope (g)
m final mass of radioisotope (g)
n number of half-lives passed (1)

$m = \dfrac{1}{2^n} m_0$

$m = \dfrac{1}{2^{0.5}} \times 6.2$

$m = \dfrac{1}{1.414 \ldots} \times 6.2$

$m = 4.384 \ldots$ g

4.38 g of the radioisotope will be left.

3.

m_0 initial mass of radioisotope (g)
m final mass of radioisotope (g)
n number of half-lives passed (1)

$$m = \frac{1}{2^n} m_0$$

$$2^n = \frac{m_0}{m}$$

$$n = \log_2 \frac{m_0}{m}$$

$$n = \log_2 \frac{20.6}{12.7}$$

$$n = \log_2 1.622$$

$$n = 0.697 \ldots \text{ half-lives}$$

0.70 half-lives have passed.

4.

m_0 initial mass of radioisotope (g)
m final mass of radioisotope (g)
n number of half-lives passed (1)

$$m = \frac{1}{2^n} m_0$$

$$2^n = \frac{m_0}{m}$$

$$n = \log_2 \frac{m_0}{m}$$

$$n = \log_2 \frac{138}{4.5}$$

$$n = \log_2 30.666 \ldots$$

$$n = 4.938 \ldots \text{ half-lives}$$

4.9 half-lives will have to pass.

5.

m_0 initial mass of radioisotope (g)
m final mass of radioisotope (g)
n number of half-lives passed (1)

$$m = \frac{1}{2^n} m_0$$

$$\frac{1}{2^n} m_0 = m$$

$$m_0 = 2^n m$$

$$m_0 = 2^{3.8} \times 0.8$$

$$m_0 = 13.928 \ldots \times 0.8$$

$$m_0 = 11.143 \ldots \text{ g}$$

We started with 11.1 g of the radioisotope.

6.

m_0 initial mass of radioisotope (g)
m final mass of radioisotope (g)
n number of half-lives passed (1)

$$m = \frac{1}{2^n} m_0$$

$$\frac{1}{2^n} m_0 = m$$

$$m_0 = 2^n m$$

$$m_0 = 2^{0.75} \times 12.3$$

$$m_0 = 1.681 \ldots \times 12.3$$

$$m_0 = 20.686 \ldots \text{ g}$$

We started with 20.7 g of the radioisotope.

7.

t time passed (days)
t_h half-life of the radioisotope (days)
n number of half-lives passed (1)

$$n = \frac{t}{t_h}$$

$$n = \frac{12.5}{5.7}$$

$$n = 2.192 \ldots \text{ half-lives}$$

Time passed corresponds to the passage of 2.19 half-lives.

8.

t time passed (years)
t_h half-life of the radioisotope (years)
n number of half-lives passed (1)

$$n = \frac{t}{t_h}$$

$$n = \frac{0.5}{4.5}$$

$$n = 0.111 \ldots \text{ half-lives}$$

Time passed corresponds to the passage of 0.11 half-lives.

9.

t time passed (s)
t_h half-life of the radioisotope (s)
n number of half-lives passed (1)

$$n = \frac{t}{t_h}$$

$$\frac{t}{t_h} = n$$

$$t = t_h \times n$$

$$t = 9.3 \times 7.7$$

$$t = 71.61 \text{ s}$$

The number of half-lives corresponds to the passage of 71.6 s.

10.

t time passed (min)
t_h half-life of the radioisotope (min)
n number of half-lives passed (1)

$$n = \frac{t}{t_h}$$

$$\frac{t}{t_h} = n$$

$$t = t_h \times n$$

$$t = 0.9 \times 12.8$$

$$t = 11.52 \text{ min}$$

The number of half-lives corresponds to the passage of 11.5 min.

11.

t time passed (min)
t_h half-life of the radioisotope (min)
n number of half-lives passed (1)

$$n = \frac{t}{t_h}$$

$$t_h = \frac{t}{n}$$

$$t_h = \frac{13.7}{4.5}$$

$$t_h = 3.044 \ldots \text{ min}$$

The radioisotope has a half-life of 3.04 min.

12.

t time passed (days)
t_h half-life of the radioisotope (days)
n number of half-lives passed (1)

$$n = \frac{t}{t_h}$$

$$t_h = \frac{t}{n}$$

$$t_h = \frac{9.2}{0.8}$$

$$t_h = 11.5 \text{ days}$$

The radioisotope has a half-life of 11.5 days.

13.

m_0 initial mass of ^{24}Na (g)
m final mass of ^{24}Na (g)
t time passed (days)
t_h half-life of ^{24}Na (days)

$$m = \frac{1}{2^{\frac{t}{t_h}}} m_0$$

$$m = \frac{1}{2^{\frac{48}{15}}} \times 7.8$$

$$m = \frac{1}{2^{3.2}} \times 7.8$$

$$m = \frac{1}{9.189\ldots} \times 7.8$$

$$m = 0.848\ldots$$

0.85 g of ^{24}Na will be left.

14.

m_0 initial mass of ^{32}P (g)
m final mass of ^{32}P (g)
t time passed (days)
t_h half-life of ^{32}P (days)

$$m = \frac{1}{2^{\frac{t}{t_h}}} m_0$$

$$m = \frac{1}{2^{\frac{2}{14.2}}} \times 28.7$$

$$m = \frac{1}{2^{0.140\ldots}} \times 28.7$$

$$m = \frac{1}{1.102\ldots} \times 28.7$$

$$m = 26.030\ldots$$

26.0 g of ^{32}P will be left.

15.

m_0 initial mass of ^{134}Cs (kg)
m final mass of ^{134}Cs (kg)
t time passed (years)
t_h half-life of ^{134}Cs (years)

$$m = \frac{1}{2^{\frac{t}{t_h}}} m_0$$

$$2^{\frac{t}{t_h}} = \frac{m_0}{m}$$

$$\frac{t}{t_h} = \log_2 \frac{m_0}{m}$$

$$t = t_h \log_2 \frac{m_0}{m}$$

$$t = 2.06 \times \log_2 \frac{132}{24.54}$$

$$t = 2.06 \times \log_2 5.378 \ldots$$

$$t = 2.06 \times 2.427 \ldots$$

$$t = 5.000 \ldots \text{ years}$$

The sample has been in storage for 5.00 years.

16.

m_0 initial mass of ^{241}Am (g)
m final mass of ^{241}Am (g)
t time passed (years)
t_h half-life of ^{241}Am (years)

$$m = \frac{1}{2^{\frac{t}{t_h}}} m_0$$

$$2^{\frac{t}{t_h}} = \frac{m_0}{m}$$

$$\frac{t}{t_h} = \log_2 \frac{m_0}{m}$$

$$t = t_h \log_2 \frac{m_0}{m}$$

$$t = 432 \times \log_2 \frac{0.5}{0.49}$$

$$t = 432 \times \log_2 1.020 \ldots$$

$$t = 432 \times 0.029 \ldots$$

$$t = 12.591 \ldots \text{ years}$$

It will take 12.6 years.

17.

m_0 initial mass of ^{7}Be (g)
m final mass of ^{7}Be (g)
t time passed (days)
t_h half-life of ^{7}Be (days)

$$m = \frac{1}{2^{\frac{t}{t_h}}} m_0$$

$$2^{\frac{t}{t_h}} = \frac{m_0}{m}$$

$$\frac{t}{t_h} = \log_2 \frac{m_0}{m}$$

$$t_h = \frac{t}{\log_2 \frac{m_0}{m}}$$

$$t_h = \frac{90}{\log_2 \frac{14.5}{4.47}}$$

$$t_h = \frac{90}{\log_2 3.243\ldots}$$

$$t_h = \frac{90}{1.697\ldots}$$

$$t_h = 53.012 \ldots \text{ days}$$

^{7}Be has a half-life of 53 days.

18.

m_0 initial mass of ^{99m}Tc (g)
m final mass of ^{99m}Tc (g)
t time passed (h)
t_h half-life of ^{99m}Tc (h)

$$m = \frac{1}{2^{\frac{t}{t_h}}} m_0$$

$$2^{\frac{t}{t_h}} = \frac{m_0}{m}$$

$$\frac{t}{t_h} = \log_2 \frac{m_0}{m}$$

$$t_h = \frac{t}{\log_2 \frac{m_0}{m}}$$

$$t_h = \frac{9}{\log_2 \frac{14.8}{5.23}}$$

$$t_h = \frac{9}{\log_2 2.829 \ldots}$$

$$t_h = \frac{9}{1.500 \ldots}$$

$$t_h = 5.997 \ldots \text{ h}$$

^{99m}Tc has a half-life of 6.00 h.

19.

m_0 initial mass of ^{22}Na (g)
m final mass of ^{22}Na (g)
t time passed (years)
t_h half-life of ^{22}Na (years)

$$m = \frac{1}{2^{\frac{t}{t_h}}} m_0$$

$$\frac{1}{2^{\frac{t}{t_h}}} m_0 = m$$

$$m_0 = 2^{\frac{t}{t_h}} m$$

$$m_0 = 2^{\frac{0.25}{2.6}} \times 12.8$$

$$m_0 = 2^{0.096 \ldots} \times 12.8$$

$$m_0 = 1.068 \ldots \times 12.8$$

$$m_0 = 13.682 \ldots \text{ g}$$

We started with 13.7 g of ^{22}Na.

20.

m_0 initial mass of ^{66}Ga (g)
m final mass of ^{66}Ga (g)
t time passed (h)
t_h half-life of ^{66}Ga (h)

$$m = \frac{1}{2^{\frac{t}{t_h}}} m_0$$

$$\frac{1}{2^{\frac{t}{t_h}}} m_0 = m$$

$$m_0 = 2^{\frac{t}{t_h}} m$$

$$m_0 = 2^{\frac{24}{9.4}} \times 2.69$$

$$m_0 = 2^{2.553\cdots} \times 2.69$$

$$m_0 = 5.869\ldots \times 12.8$$

$$m_0 = 75.127\ldots \text{ g}$$

We started with 75.1 g of ^{66}Ga.

21.

r percentage of the mass of radioisotope left (1)
n number of half-lives passed (1)

$$r = \frac{1}{2^n}$$

$$r = \frac{1}{2^{1.75}}$$

$$r = \frac{1}{3.363\ldots}$$

$$r = 0.297\ldots$$

29.7% of the mass of the radioisotope will be left.

22.

r percentage of the mass of radioisotope left (1)
n number of half-lives passed (1)

$$r = \frac{1}{2^n}$$

$$r = \frac{1}{2^{12.7}}$$

$$r = \frac{1}{6653.971\ldots}$$

$$r = 0.000\,150\ldots$$

0.015% of the mass of the radioisotope will be left.

23.

r percentage of the mass of radioisotope left (1)
n number of half-lives needed (1)

$$r = \frac{1}{2^n}$$

$$2^n = \frac{1}{r}$$

$$n = \log_2 \frac{1}{r}$$

$$n = \log_2 \frac{1}{0.3}$$

$$n = \log_2 3.333 \ldots$$

$$n = 1.736 \ldots$$

It takes 1.74 half-lives.

24.

r percentage of the mass of radioisotope left (1)
n number of half-lives needed (1)

$$r = \frac{1}{2^n}$$

$$2^n = \frac{1}{r}$$

$$n = \log_2 \frac{1}{r}$$

$$n = \log_2 \frac{1}{0.1}$$

$$n = \log_2 10$$

$$n = 3.321 \ldots$$

It takes 3.32 half-lives.

25.

r percentage of the mass of radioisotope left (1)
t time passed (months)
t_h half-life of radioisotope (months)

$$r = \frac{1}{2^{\frac{t}{t_h}}}$$

$$r = \frac{1}{2^{\frac{8.5}{3.2}}}$$

$$r = \frac{1}{2^{2.65625}}$$

$$r = \frac{1}{6.303 \ldots}$$

$$r = 0.158 \ldots$$

15.9% of the mass of the radioisotope will be left.

26.

r percentage of the mass of radioisotope left (1)
t time passed (h)
t_h half-life of radioisotope (h)

$$r = \frac{1}{2^{\frac{t}{t_h}}}$$

$$r = \frac{1}{2^{\frac{12.9}{15}}}$$

$$r = \frac{1}{2^{0.86}}$$

$$r = \frac{1}{1.815\ldots}$$

$$r = 0.550\ldots$$

55.1% of the mass of the radioisotope is left.

27.

r percentage of the mass of radioisotope left (1)
t time passed (days)
t_h half-life of radioisotope (days)

$$r = \frac{1}{2^{\frac{t}{t_h}}}$$

$$2^{\frac{t}{t_h}} = \frac{1}{r}$$

$$\frac{t}{t_h} = \log_2 \frac{1}{r}$$

$$t = t_h \log_2 \frac{1}{r}$$

$$t = 9.1 \log_2 \frac{1}{0.4}$$

$$t = 9.1 \log_2 2.5$$

$$t = 9.1 \times 1.321\ldots$$

$$t = 12.029\ldots \text{ days}$$

It will take 12.0 days.

28.

r percentage of the mass of radioisotope left (1)
t time passed (s)
t_h half-life of radioisotope (s)

$$r = \frac{1}{2^{\frac{t}{t_h}}}$$

$$2^{\frac{t}{t_h}} = \frac{1}{r}$$

$$\frac{t}{t_h} = \log_2 \frac{1}{r}$$

$$t = t_h \log_2 \frac{1}{r}$$

$$t = 132 \log_2 \frac{1}{0.9}$$

$$t = 132 \log_2 1.111 \ldots$$

$$t = 132 \times 0.152 \ldots$$

$$t = 20.064 \ldots \text{ s}$$

It will take 20.1 s.

29.

r percentage of the mass of radioisotope left (1)
t time passed (h)
t_h half-life of radioisotope (h)

$$r = \frac{1}{2^{\frac{t}{t_h}}}$$

$$2^{\frac{t}{t_h}} = \frac{1}{r}$$

$$\frac{t}{t_h} = \log_2 \frac{1}{r}$$

$$t_h = \frac{t}{\log_2 \frac{1}{r}}$$

$$t_h = \frac{5.5}{\log_2 \frac{1}{0.15}}$$

$$t_h = \frac{5.5}{\log_2 6.666 \ldots}$$

$$t_h = \frac{5.5}{2.736 \ldots}$$

$$t_h = 2.009 \ldots \text{ h}$$

The radioisotope has a half-life of 2.01 h.

30.

r percentage of the mass of radioisotope left (1)
t time passed (days)
t_h half-life of radioisotope (days)

$$r = \frac{1}{2^{\frac{t}{t_h}}}$$

$$2^{\frac{t}{t_h}} = \frac{1}{r}$$

$$\frac{t}{t_h} = \log_2 \frac{1}{r}$$

$$t_h = \frac{t}{\log_2 \frac{1}{r}}$$

$$t_h = \frac{3.8}{\log_2 \frac{1}{0.218}}$$

$$t_h = \frac{3.8}{\log_2 4.587 \ldots}$$

$$t_h = \frac{3.8}{2.197 \ldots}$$

$$t_h = 1.729 \ldots \text{ days}$$

The radioisotope has a half-life of 1.73 days.

31.

A activity of the radioactive substance (Bq)
N number of disintegrations (1)
t time (s)

$$A = \frac{N}{t}$$

$$A = \frac{126}{14.5}$$

$$A = 8.689 \ldots \text{ Bq}$$

The radioactive substance has an activity of 8.69 Bq.

32.

A activity of the radioactive substance (Bq)
N number of disintegrations (1)
t time (s)

$$A = \frac{N}{t}$$

$$\frac{N}{t} = A$$

$$N = At$$

$$N = 63 \times 38.5$$

$$N = 2425.5$$

The radioactive substance experiences 2426 disintegrations during this time.

33.

D absorbed dose (Gy)
E heat generated by absorbed radiation (J)
m mass of target tissue (kg)

$$D = \frac{E}{m}$$

$$D = \frac{0.11}{0.38}$$

$$D = 0.289 \ldots \text{ Gy}$$

The tissue experiences an absorbed dose of 0.289 Gy.

34.

D absorbed dose (Gy)
E heat generated by absorbed radiation (J)
m mass of target tissue (kg)

$$D = \frac{E}{m}$$

$$\frac{E}{m} = D$$

$$E = Dm$$

$$E = 0.0175 \times 0.042$$

$$E = 0.000\,735 \text{ J}$$

0.000 735 J of heat was absorbed by the tissue.

35.

H equivalent dose (Sv)
D absorbed dose (Gy)
w_R radiation weighting factor (1)

$H = w_R D$

$H = 1 \times 0.43$

$H = 0.43 \text{ Sv}$

The tissue received an equivalent dose of 0.43 Sv.

36.

H equivalent dose (Sv)
D absorbed dose (Gy)
w_R radiation weighting factor (1)

$H = w_R D$

$w_R D = H$

$w_R = \dfrac{H}{D}$

$w_R = \dfrac{0.015}{0.000\,75}$

$w_R = 20$

Alpha rays have a radiation weighing factor of 20.

37.

H_T effective dose (Sv)
H equivalent dose (Sv)
w_T tissue weighting factor (1)

$H_T = w_T H$

$H_T = 0.05 \times 0.46$

$H_T = 0.023 \text{ Sv}$

The tissue received an effective dose of 0.023 Sv.

38.

H_T effective dose (Sv)
H equivalent dose (Sv)
w_T tissue weighting factor (1)

$H_T = w_T H$

$w_T H = H_T$

$H = \dfrac{H_T}{w_T}$

$H = \dfrac{0.002\,83}{0.12}$

$H = 0.023\,583 \ldots$

The tissue received an equivalent dose of 0.0236 Sv.

Exercise Set 13

1. In each case we list the subexpression that can be turned into a constant followed by the expression with the constant in place of the subexpression.

   a.

$q_2 q_1$	$k - q_4 q_3$
$q_4 q_3$	$q_2 q_1 - k$
$q_2 q_1 - q_4 q_3$	k
q_2	$k q_1 - q_4 q_3$
q_1	$k q_2 - q_4 q_3$
q_4	$q_2 q_1 - k q_3$
q_3	$q_2 q_1 - k q_4$

   b.

q_1	$k - \dfrac{q_2}{q_3 q_4} - q_6 q_5$
$\dfrac{q_2}{q_3 q_4}$	$q_1 - k - q_6 q_5$
$q_6 q_5$	$q_1 - \dfrac{q_2}{q_3 q_4} - k$
$q_1 - \dfrac{q_2}{q_3 q_4}$	$k - q_6 q_5$
$q_1 - q_6 q_5$	$k - \dfrac{q_2}{q_3 q_4}$
$-\dfrac{q_2}{q_3 q_4} - q_6 q_5$	$q_1 + k$
$q_1 - \dfrac{q_2}{q_3 q_4} - q_6 q_5$	k
q_2	$q_1 - \dfrac{k}{q_3 q_4} - q_6 q_5$
q_3	$q_1 - k \dfrac{q_2}{q_4} - q_6 q_5$
q_4	$q_1 - k \dfrac{q_2}{q_3} - q_6 q_5$
$\dfrac{q_2}{q_3}$	$q_1 - \dfrac{k}{q_4} - q_6 q_5$
$\dfrac{q_2}{q_4}$	$q_1 - \dfrac{k}{q_3} - q_6 q_5$
$\dfrac{1}{q_3 q_4}$	$q_1 - k q_2 - q_6 q_5$
q_6	$q_1 - \dfrac{q_2}{q_3 q_4} - k q_5$
q_5	$q_1 - \dfrac{q_2}{q_3 q_4} - k q_6$

2. The solutions below list the quantity symbols that we wish to relate, followed by a rewriting of the formula to incorporate constants with a note on what the constant stands for. Every now and then, we make brief comments about the semantics behind the relationships.

a.
- V and I: $V = kI$, $k = R$
  direct proportion

- V and R: $V = kR$, $k = I$
  direct proportion

- I and R: $I = \frac{k}{R}$, $k = V$
  inverse proportion

This formula relates the voltage across a resistor to the current through the resistor and the resistance of the resistor. Voltage relates to the force that drives free electrons through the resistor while current refers to the number of electrons that pass through the resistor per unit time. Resistance refers to the property of the resistor that tend to prevent the free electrons flowing through it.

With these definitions in mind, the second relationship above states that the voltage across a resistor is directly proportional to the resistance of the resistor. As an example, to drive the same current through a resistor that has twice the resistance of another resister, we need to set up a voltage across it that is twice larger than the one across the other resistor.

b.
- T and f: $T = \frac{k}{f}$, $k = 1$
  inverse proportion

This formula relates the period of a cyclic event to the number of cycles that repeat per unit time.

The relationship above states that the period of a cyclic event is inversely proportional to its period. As an example, if 2 cycles occur in 1 s (frequency is 2 cycles/s), then each cycle takes $\frac{1}{2}$ s (period is $\frac{1}{2}$ s) and if 3 cycles occur in 1 s (frequency is 3 cycles/s), then each cycle takes $\frac{1}{3}$ s (period is $\frac{1}{3}$ s) and so on.

c.
- A and b: $A = kb$, $k = \frac{1}{2}h$
  direct proportion

- A and h: $A = kh$, $k = \frac{1}{2}b$
  direct proportion

- b and h: $b = \frac{k}{h}$, $k = 2A$
  inverse proportion

This formula relates the area of a triangle to the lengths of its base and height.

The semantics behind these relationships are interesting. As an example, the second relationship states that doubling the height of a triangle doubles its area or that reducing the height of a rectangle to $\frac{1}{3}$ of its initial size reduces the area of the triangle by $\frac{1}{3}$. As a second example of how a knowledge of the type of relationship between quantities allows us to make conclusions, the last relationship tells us that if we double the height of a triangle but wish to keep its area the same, then we will have to halve its base.

d.   • p and F: $p = kF$, $k = \frac{1}{A}$
      direct proportion

      • p and A: $p = \frac{k}{A}$, $k = F$
      inverse proportion

This formula relates the pressure over a surface to the uniform force acting on that surface and the area of the surface.

We let the reader make sense of the relationships above.

e.   • E and m: $E = km$, $k = \frac{1}{2}v^2$
      direct proportion

      • E and v: $E = kv^2$, $k = \frac{1}{2}m$
      nonlinear but not inverse proportion

      • m and v: $m = \frac{k}{v^2}$, $k = 2E$
      nonlinear but not inverse proportion

This formula relates the kinetic energy of a moving object to its mass and its speed.

Note that the second relationship is not linear as v does not form a factor on its own (It would have been direct proportion if the equation had turned into $E = kv$, not $E = kv^2$).

f.   • m and n: $m = \frac{k}{2^n}$, $k = m_0$
      nonlinear but not inverse proportion

      • m and m_0: $m = km_0$, $k = \frac{1}{2^n}$
      direct proportion

      • n and m_0: $m_0 = 2^n k$, $k = m$
      nonlinear but not inverse proportion

This is the radiation equation that has been discussed in detail in the body of the textbook.

g.   • p_1 and V_1: $p_1 = \frac{k}{V_1}$, $k = p_2 V_2$
      inverse proportion

      • p_1 and p_2: $p_1 = kp_2$, $k = \frac{V_2}{V_1}$
      direct proportion

      • p_1 and V_2: $p_1 = kV_2$, $k = \frac{p_2}{V_1}$
      direct proportion

      • V_1 and p_2: $V_1 = kp_2$, $k = \frac{V_2}{p_1}$
      direct proportion

      • V_1 and V_2: $V_1 = kV_2$, $k = \frac{p_2}{p_1}$
      direct proportion

- p_2 and V_2: $p_2 = \frac{k}{V_2}$, $k = p_1 V_1$
  inverse proportion

This formula relates the pressure of an ideal gas to the volume of its container and can be derived from the Ideal Gas Equation of State if we keep the amount of the gas and its temperature constant. The equation is based on the conservation form of the equation for inverse proportion.

h.
- p_1 and T_1: $p_1 = kT_1$, $k = \frac{p_2}{T_2}$
  direct proportion

- p_1 and p_2: $p_1 = kp_2$, $k = \frac{T_1}{T_2}$
  direct proportion

- p_1 and T_2: $p_1 = \frac{k}{T_2}$, $k = T_1 p_2$
  inverse proportion

- T_1 and p_2: $T_1 = \frac{k}{p_2}$, $k = T_2 p_1$
  inverse proportion

- T_1 and T_2: $T_1 = kT_2$, $k = \frac{p_1}{p_2}$
  direct proportion

- p_2 and T_2: $p_2 = kT_2$, $k = \frac{p_1}{T_1}$
  direct proportion

This formula relates the pressure of an ideal gas to its temperature and can be derived from the Ideal Gas Equation of State if we keep the volume of the container and the amount of the gas constant. The equation is based on the conservation form of the equation for direct proportion.

i.
- F and m: $F = km$, $k = \frac{v^2}{r}$
  direct proportion

- F and v: $F = kv^2$, $k = \frac{m}{r}$
  nonlinear but not inverse proportion

- F and r: $F = \frac{k}{r}$, $k = mv^2$
  inverse proportion

- m and r: $m = kr$, $k = \frac{F}{v^2}$
  direct proportion

- m and v: $m = \frac{k}{v^2}$, $k = rF$
  nonlinear but not inverse proportion

- r and v: $r = kv^2$, $k = \frac{m}{F}$

nonlinear but not inverse proportion

This formula relates the force acting on an object moving along a circular path to the mass of the object, the speed of the object and the radius of the circular path.

We let the reader make a few semantic conclusions based on the relationships above.

j.   • P and l: $P = ml + b$, $m = 2$, $b = 2w$
         linear

     • P and w: $P = mw + b$, $m = 2$, $b = 2l$
         linear

     • l and w: $l = -w + \frac{1}{2}P$, $m = -1$, $b = \frac{1}{2}P$
         linear

This formula relates the perimeter, P, of a rectangle to its length, l, and width, w.

3. a. Since V is inversely proportional to p, scaling p by a factor of 4 scales V by a factor of $\frac{1}{4}$.

   Since V is directly proportional to T, scaling T by a factor of 8 scales V by a factor of 8.

   The combined effect of changes in both p and T will scale V by a factor of $\frac{1}{4} \times 8$ or 2, i.e., V doubles.

   b. Since T is directly proportional to V, scaling V by a factor of 2 scales T by a factor of 2.

   Since T is inversely proportional to n, scaling n by a factor of $\frac{1}{3}$ scales T by a factor of 3.

   The combined effect of changes in both V and n is to scale T by a factor of 2×3 or 6.

   c. V scales by a factor of 36.

   We let the reader work out the detail.

   d. n scales by a factor of 3.

   We let the reader work out the detail.

4. a. Since d is directly superpositional to d_0, an increase in d_0 by 12 m increases d by 12 m.

   Since d is directly superpositional to vt, a decrease in vt by 5 m decreases d by 5 m.

   The combined effect of changes in both d_0 and vt is to increase the value of d by $12 - 5$ or 7 units.

   b. The covered distance, vt, would have to increase by 40 m.

   We let the reader explain why.

   The change can be achieved by driving faster, extending the final time, or both, in ways that would raise the value of vt by 40 m.

c. Since vt is directly proportional to v, with t constant, scaling vt by a factor of 2 implies scaling v by a factor of 2, i.e., the speed will have to double.

Exercise Set A

1. Quantity name: mass
   Quantity symbol: m
   Entity: CO_2
   Value of the quantity: 45.5 g
   Numerical value of the quantity: 45.5
   Unit name: gram
   Unit symbol: g

2. Quantity name: cost
   Quantity symbol: c
   Entity: book
   Value of the quantity: $9.20
   Numerical value of the quantity: 9.20
   Unit name: dollar
   Unit symbol: $

3. Quantity name: rate
   Quantity symbol: r
   Entity: discount
   Value of the quantity:[16] 15% 1
   Numerical value of the quantity: 15%
   Unit name: one
   Unit symbol: 1

4. Quantity name: speed
   Quantity symbol: s or v
   Entity: train
   Value of the quantity: 200 km/h
   Numerical value of the quantity: 200
   Unit name: kilometres per hour
   Unit symbol: km/h

5. Quantity name: area
   Quantity symbol: A
   Entity: lot
   Value of the quantity: 4000 ft^2
   Numerical value of the quantity: 4000
   Unit name: square feet
   Unit symbol: ft^2

6. Quantity name: time
   Quantity symbol: t
   Entity: operation
   Value of the quantity: 1.5 h
   Numerical value of the quantity: 1.5
   Unit name: hour
   Unit symbol: h

7. Quantity name: temperature
   Quantity symbol: t
   Entity: sun
   Value of the quantity: 6000 °C
   Numerical value of the quantity: 6000
   Unit name: degree Celsius
   Unit symbol: °C

8. Quantity name: energy
   Quantity symbol: E
   Entity: bullet
   Value of the quantity: 2025 J
   Numerical value of the quantity: 2025
   Unit name: joule
   Unit symbol: J

9. Quantity name: acceleration
   Quantity symbol: a
   Entity: BMW
   Value of the quantity: 12.66 m/s^2
   Numerical value of the quantity: 12.66
   Unit name: metres per square second
   Unit symbol: m/s^2

10. Quantity name: mass
    Quantity symbol: m
    Entity: Charlie
    Value of the quantity: 32 kg
    Numerical value of the quantity: 32
    Unit name: kilogram
    Unit symbol: kg

[16]The notation 15% 1 implies 15% of 1. For a detailed account of the use of *one* as the unit associated with percentage, please see Chapter 7.

[17]For a detailed account of the difference between the definitions of mass and weight in the sciences and the difference between weight as defined in the sciences and weight in its ordinary sense please see the chapter on measurement in the companion textbook *Tactical Style in Mathematics* by the author.

11. Quantity name:[17] weight   Numerical value of the quantity: 314
      Quantity symbol: W Unit name: newton
      Entity: Charlie   Unit symbol: N
      Value of the quantity: 314 N

Exercise Set B.1

1. An atom is the smallest unit of matter that retains its chemical properties.[18]

2. a. O   c. Ar   e. Zn   g. C
   b. Mg   d. S   f. Cl   h. Ca

3. a. nitrogen   c. fluorine   e. iron   g. carbon
   b. neon   d. sodium   f. lead   h. iodine

4. Atoms within the same column in the Periodic Table have similar chemical properties.

5. a. beryllium   c. oxygen   e. potassium   g. cadmium
   b. aluminum   d. neon   f. helium   h. lithium

Exercise Set B.2

1. protons, neutrons and electrons

2. a. 1 amu
   b. approximately 1 amu
   c. approximately 0.0005 amu

3. a. +1 e   b. 0 e   c. −1 e

4. A tiny volume at the centre of the atom in which the protons and neutrons are packed.

5. The number of protons in the nucleus of an atom.

6. The number of protons in the nucleus of an atom sets the type of the atom. Therefore, a knowledge of the number of protons in the nucleus of an atom tells us what atom we are dealing with.

7. In the Periodic Table, the atomic number is written on top of the chemical symbol for that atom within the cell for that atom.

8. In reactions, the atomic number is written as a subscript on the bottom, left side of the symbol for the atom.

---

[18]By **chemical properties** we mean the manner in which it interacts with other atoms and molecules (groupings of atoms).

9. a. 11   c. 35   e. 5   g. 8
   b. 79   d. 7   f. 2   h. 16

10. Atoms of the same kind (atoms that have the same number of protons or the same atomic number) that have different number of neutrons (different mass numbers).[19]

11. The number of protons and neutrons in the nucleus of an atom.

12. The adjective *mass* in *mass number* is rooted in the fact that the numerical value of the total *number* of protons and neutrons in the nucleus of an isotope is equal to the numerical value of the *mass* of that isotope in amu.

13. In reactions the mass number is written as a superscript on the top, left side of the symbol for the atom.

14. a. mass number:[20] 17 1; mass: 17 amu
    b. mass number: 35 1; mass: 35 amu
    c. mass number: 99 1; mass: 99 amu
    d. mass number: 131 1; mass: 131 amu

15. In stable isotopes of most atoms the number of neutrons is usually equal, or otherwise close, to the number of protons in their nuclei.

16. a. I-131, iodine-131   b. ^{35}S, sulfur-35   c. ^{3}H, H-3

17. a. ^{17}O has 17 protons and neutrons.
       From the Periodic Table, an O atom has 8 protons.
       Therefore, ^{17}O has $17 - 8$ or 9 neutrons.
    b. ^{35}S has 35 protons and neutrons.
       From the Periodic Table, an S atom has 16 protons.
       Therefore, ^{35}S has $35 - 16$ or 19 neutrons.
    c. ^{99}Tc has 99 protons and neutrons.
       From the Periodic Table, a Tc atom has 43 protons.
       Therefore, ^{99}Tc has $99 - 43$ or 56 neutrons.
    d. ^{131}I has 131 protons and neutrons.
       From the Periodic Table, an I atom has 53 protons.
       Therefore, ^{131}I has $131 - 53$ or 78 neutrons.
    e. ^{19}F has 19 protons and neutrons.
       From the Periodic Table, an F atom has 9 protons.
       Therefore, ^{19}F has $19 - 9$ or 10 neutrons.
    f. ^{3}H has 3 protons and neutrons.
       From the Periodic Table, an H atom has 1 proton.
       Therefore, ^{3}H has $3 - 1$ or 2 neutrons.

---

[19]Since the number of protons are the same, the difference in mass number comes from the difference in the number of neutrons.

[20]The unit symbol 1 in the value of the mass numbers in this solution refers to the number of protons and neutrons. Therefore, 17 1 should be interpreted as 17 protons and neutrons.

g. ^{186}U has 186 protons and neutrons.
From the Periodic Table, a U atom has 92 protons.
Therefore, ^{186}U has 186 − 92 or 94 neutrons.

h. ^{31}P has 31 protons and neutrons.
From the Periodic Table, a P atom has 15 protons.
Therefore, ^{31}P has 31 − 15 or 16 neutrons.

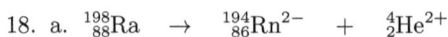

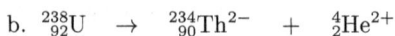

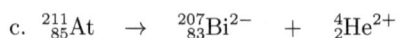

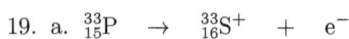

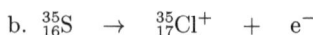

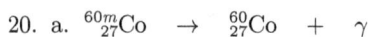

18. a. $^{198}_{88}\text{Ra}$ → $^{194}_{86}\text{Rn}^{2-}$ + $^{4}_{2}\text{He}^{2+}$

b. $^{238}_{92}\text{U}$ → $^{234}_{90}\text{Th}^{2-}$ + $^{4}_{2}\text{He}^{2+}$

c. $^{211}_{85}\text{At}$ → $^{207}_{83}\text{Bi}^{2-}$ + $^{4}_{2}\text{He}^{2+}$

19. a. $^{33}_{15}\text{P}$ → $^{33}_{16}\text{S}^{+}$ + e^{-}

b. $^{35}_{16}\text{S}$ → $^{35}_{17}\text{Cl}^{+}$ + e^{-}

c. $^{3}_{1}\text{H}$ → $^{3}_{2}\text{He}^{+}$ + e^{-}

20. a. $^{60m}_{27}\text{Co}$ → $^{60}_{27}\text{Co}$ + γ

b. $^{65m}_{30}\text{Zn}$ → $^{65}_{30}\text{Zn}$ + γ

c. $^{137m}_{55}\text{Cs}$ → $^{137}_{55}\text{Cs}$ + γ

21. Atomic mass is the average mass of the isotopes of an atom from a sample of
that atom from nature.

22. a. 16.00 amu/atom
    b. 32.07 amu/atom
    c. Technetium does not have isotopes with characteristic isotopic abundance in
       natural terrestrial samples.
    d. 126.9 amu/atom
    e. 19.00 amu/atom
    f. 1.008 amu/atom
    g. 238.0 amu/atom
    h. 30.97 amu/atom

23. The number of electrons in an atom is equal to the number of protons in that
atom's nucleus.

24. a. 8   c. 43   e. 9   g. 92
    b. 16   d. 53   f. 1   h. 15

25. An ion is an atom that has gained extra electrons or has lost electrons. An anion
is an atom that has gained extra electrons and, as such, carries a net negative
charge. A cation is an atom that has lost electrons and, as such, carries a net
positive charge.

26. a. F^- b. O^{2-} c. Li^+ d. Al^{3+}

27. The number of electrons in an ion does not equal to the number of protons in that ion's nucleus.

Exercise Set B.3

1. A covalent bond is a bond that forms when atoms with tendencies to gain electrons share electrons to reach a more stable state. A nonpolar covalent bond forms when the two atoms forming the covalent bond have similar tendencies to attract electrons. In such cases the electrons are shared equally or almost equally giving rise to a bond that is not polar. A polar covalent bond forms when one of the atoms involved has a noticeably higher tendency to attract electrons compared to the other. In such cases the bond will be polar as the shared electrons will be pulled towards the atom with the stronger tendency to attract electrons, making its side slightly negative and the other side slightly positive.

2. A molecule is a grouping of atoms that form a unit separate and distinct from other atoms and molecules.

3. a. 1 atom of C, 4 atoms of H
   b. 2 atoms of C, 6 atoms of H
   c. 1 atom of S, 3 atoms of O
   d. 2 atoms of N, 4 atoms of O
   e. 6 atoms of C, 12 atoms of H, 6 atoms of O
   f. 3 atoms of O
   g. 2 atoms of Cl
   h. 1 atom of C, 2 atoms of S

4. An ionic bond forms when an atom with a tendency to gain electrons nears an atom with a tendency to lose electrons. The combined tendencies result in the transfer of one or more electrons from the atom with the tendency to lose electrons to the atom with the tendency to gain electrons, resulting in the formation of oppositely charged ions that form a bond due to the force of attraction between them.

5. An ionic substance is made of many ions that stick to each other forming crystal-like structures.

6. a. Ions of Mg^{2+} and ions of Cl^-.
   b. Ions of Na^+ and ions of I^-.
   c. Ions of Na^+ and ions of S^{2-}.

7. Metals are atoms with a tendency to lose electrons. Nonmetals are atoms with a tendency to gain electrons. Metalloids occupy a gray area between metals and nonmetals.

8. a. 5×1 or 5 C atoms, 5×4 or 20 H atoms
   b. 2×2 or 4 C atoms, 2×6 or 12 H atoms
   c. 8×1 or 8 S atoms, 8×3 or 24 O atoms
   d. 12×2 or 24 N atoms, 12×4 or 48 O atoms
   e. 15×6 or 90 C atoms, 15×12 or 180 H atoms, 15×6 or 90 O atoms

   f. 2×3 or 6 O atoms

   g. 3×2 or 6 Cl atoms

   h. 95×1 or 95 C atoms, 95×2 or 190 S atoms

Exercise Set B.4

1. A physical process is a process that results in observable changes in the state of a substance without affecting the composition of the molecules or the types of atoms within the molecules that make up the substance.

   A chemical reaction is a process that results in changes in the molecular composition of a substance, i.e., the grouping of the atoms within the moleculs, but leaves the atoms themselves unchanged.

   A nuclear reaction is a process that results in changes in the atoms themselves.

2. No. Physical changes leave molecular composition unchanged.

   Yes. By definition a chemical reaction results in changes in the composition of molecules that make up the substance.

   No. By definition a nuclear reaction results in changes to atoms themselves but leaves their grouping unchanged.

3. No. Physical changes leave atoms unchanged.

   No. Chemical reactions leave atoms unchanged.

   Yes. By definition a nuclear reaction results in changes in the nucleus of an atom which often results in the conversion of atoms of one kind to atoms of a different kind.

4. a. Reactants: CH_4 and O_2
      Products: CO_2 and H_2O

   b. Reactants: CO_2 and H_2O
      Products: $C_6H_{12}O_6$ and O_2

   c. Reactants: Fe and S_8
      Products: FeS

   d. Reactants: Mg and H_2O
      Products: $Mg(OH)_2$ and H_2

   e. Reactants: Fe and NaBr
      Products: $FeBr_3$ and Na

   f. Reactants: NaOH and KNO_3
      Products: $NaNO_3$ and KOH

   g. Reactants: $CaCl_2$ and $AgNO_3$
      Products: AgCl and $Ca(NO_3)_2$

   h. Reactants: H_2SO_4 and KOH
      Products: K_2SO_4 and H_2O

5. a. This chemical reaction is correctly balanced: There is 1×1 or 1 C atom on the left side and there is 1×1 or 1 C atom on the right side. There are 1×4 or 4 H atoms on the left side and 2×2 or 4 H atoms on the right side. There are 2×2 or 4 O atoms on the left side and $1 \times 2 + 2 \times 1$ or $2 + 2$ or 4 O atoms on the right side.

   b. This chemical reaction is correctly balanced: There are 6×1 or 6 C atoms on the left side and 1×6 or 6 C atoms on the right side. There are $6 \times 2 + 6 \times 1$ or $12 + 6$ or 18 O atoms on the left side and $1 \times 6 + 6 \times 2$ or $6 + 12$ or 18 O atoms on the right side. There are 6×2 or 12 H atoms on the left side and 1×12 or 12 H atoms on the right side.

c. This chemical reaction is not correctly balanced: There is 1×1 or 1 Fe atom on the left side but there are 2×1 or 2 Fe atoms on the right side.

d. This reaction is not correctly balanced: There are 3×1 or 3 O atoms on the left side but there are only $1 \times 2 \times 1$ or 2 O atoms on the right side.

e. This chemical reaction is correctly balanced: There is 1×1 or 1 Fe atom on the left side and there is 1×1 or 1 Fe atom on the right side. There are 3×1 or 3 Na atoms on the left side and 3×1 or 3 Na atoms on the right side. There are 3×1 or 3 Br atoms on the left side and 1×3 or 3 Br atoms on the right side.

f. This chemical reaction is correctly balanced: There is 1×1 or 1 Na atom on the left side and 1×1 or 1 Na atom on the right side. There are $1 \times 1 + 1 \times 3$ or $1 + 3$ or 4 O atoms on the left side and $1 \times 3 + 1 \times 1$ or $3 + 1$ or 4 O atoms on the right side. There is 1×1 or 1 H atom on the left side and 1×1 or 1 H atom on the right side. There is 1×1 or 1 K atom on the left side and 1×1 or 1 K atom on the right side. There is 1×1 or 1 N atom on the left side and 1×1 or 1 N atom on the right side.

g. This chemical reaction is not correctly balanced: There is 1×1 or 1 Ca atom on the left side but 3×1 or 3 Ca atoms on the right side.

h. This chemical reaction is correctly balanced: There are $1 \times 2 + 2 \times 1$ or $2 + 2$ or 4 H atoms on the left side and 2×2 or 4 H atoms on the right side. There is 1×1 or 1 S atom on the left side and 1×1 or 1 S atom on the right side. There are $1 \times 4 + 2 \times 1$ or $4 + 2$ or 6 O atoms on the left side and $1 \times 4 + 2 \times 1$ or $4 + 2$ or 6 O atoms on the right side.

Exercise Set B.5

1. a. A unit of mass suitable for measuring the values of the masses of small amounts of atoms or molecules is *atomic mass unit*. The reason is that atomic mass unit is a unit of mass whose size is within the range of masses of small amounts of atoms or molecules. The unit symbol for the unit *atomic mass unit* is amu.

   b. A unit of energy suitable for measuring the values of the energies of small amounts of atoms or molecules is *electron volt*. The reason is that electron volt is a unit of energy whose size is within the range of energies of small amounts of atoms or molecules. The unit symbol for the unit *electron volt* is eV.

   c. A unit of amount suitable for measuring the values of the amounts of small amounts of atoms or molecules is *one*. The reason is that one is a unit of amount whose size is within the range of amounts of small amounts of atoms or molecules. The unit symbol for the unit *one* is 1.

2. a. A unit of mass suitable for measuring the values of the masses of chunks of matter is *gram*. The reason is that gram is a unit of mass whose size is within the range of masses of chunks of matter that we work with on a daily basis. The unit symbol for the unit *gram* is g.

   b. A unit of energy suitable for measuring the values of the energies of chunks of matter is *joule*. The reason is that joule is a unit of energy whose size is within the range of energies of chunks of matter that we work with on a daily basis. The unit symbol for the unit *joule* is J.

c. A unit of amount suitable for measuring the values of the amounts of chunks of matter is *mole*. The reason is that mole is a unit of amount whose size is within the range of amounts of chunks of matter that we work with on a daily basis. The unit symbol for the unit *mole* is mol.

3. The reason we choose the amount of atoms in 12.01 g of carbon as 1 mol of carbon is that the numerical value maps onto the atomic mass of carbon in amu. This equality in the numerical values of the atomic mass of carbon and the molar mass of carbon (or, indeed, of any entity) makes it easier for us to work with individual atoms or chunks of matter by allowing interpretations of our calculations involving mass as being carried out in units of amu/atom or g/mol.

4. Atomic mass is the average mass of an atom. It is computed as the average mass of the isotopes of the atom in a sample from nature.

5. Molar mass is the mass per 1 mole of a substance. It is computed as the average mass of 1 mol of the isotopes of the substance in a sample from nature.

6. a. 16.00 g/mol
   b. 32.07 g/mol
   c. Technetium does not have isotopes with characteristic isotopic abundance in natural terrestrial samples.
   d. 126.9 g/mol
   e. 19.00 g/mol
   f. 1.008 g/mol
   g. 238.0 g/mol
   h. 30.97 g/mol

7. a. 5×2 or 10 mol of C atoms, 5×6 or 30 mol of H atoms
   b. 12×1 or 12 mol of C atoms, 12×2 or 24 mol of O atoms
   c. 4×8 or 32 mol of S atoms
   d. 3×2 or 6 mol of H atoms, 3×1 or 3 mol of O atoms
   e. 1×1 or 1 mol of Na atoms, 1×1 or 1 mol of N atoms, 1×3 or 3 mol of O atoms
   f. 15×1 or 15 mol of Ca atoms, 15×2 or 30 mol of Cl atoms
   g. 6×2 or 12 mol of H atoms, 6×1 or 6 mol of S atoms, 6×4 or 24 mol of O atoms
   h. 1×1 or 1 mol of C atoms, 1×1 or 1 mol of O atoms

www.ingramcontent.com/pod-product-compliance
Lightning Source LLC
Chambersburg PA
CBHW060031210326
41520CB00009B/1086